Springer Theses

Recognizing Outstanding Ph.D. Research

For further volumes:
http://www.springer.com/series/8790

Aims and Scope

The series "Springer Theses" brings together a selection of the very best Ph.D. theses from around the world and across the physical sciences. Nominated and endorsed by two recognized specialists, each published volume has been selected for its scientific excellence and the high impact of its contents for the pertinent field of research. For greater accessibility to non-specialists, the published versions include an extended introduction, as well as a foreword by the student's supervisor explaining the special relevance of the work for the field. As a whole, the series will provide a valuable resource both for newcomers to the research fields described, and for other scientists seeking detailed background information on special questions. Finally, it provides an accredited documentation of the valuable contributions made by today's younger generation of scientists.

Theses are accepted into the series by invited nomination only and must fulfill all of the following criteria

- They must be written in good English.
- The topic should fall within the confines of Chemistry, Physics, Earth Sciences, Engineering and related interdisciplinary fields such as Materials, Nanoscience, Chemical Engineering, Complex Systems and Biophysics.
- The work reported in the thesis must represent a significant scientific advance.
- If the thesis includes previously published material, permission to reproduce this must be gained from the respective copyright holder.
- They must have been examined and passed during the 12 months prior to nomination.
- Each thesis should include a foreword by the supervisor outlining the significance of its content.
- The theses should have a clearly defined structure including an introduction accessible to scientists not expert in that particular field.

Takayuki Yuasa

Suzaku Studies of White Dwarf Stars and the Galactic X-ray Background Emission

Doctoral Thesis accepted by
the University of Tokyo, Japan

Springer

Author
Dr. Takayuki Yuasa
Department of Physics
Graduate School of Science
The University of Tokyo
Tokyo
Japan

Supervisor
Prof. Kazuo Makishima
Department of Physics
Graduate School of Science
The University of Tokyo
Tokyo
Japan

ISSN 2190-5053 ISSN 2190-5061 (electronic)
ISBN 978-4-431-54218-6 ISBN 978-4-431-54219-3 (eBook)
DOI 10.1007/978-4-431-54219-3
Springer Tokyo Heidelberg New York Dordrecht London

Library of Congress Control Number: 2012949067

Printed on acid-free paper

Springer is part of Springer Science+Business Media (www.springer.com)

Parts of this thesis have been published in the following articles:

Takayuki Yuasa, Kazuo Makishima, Kazuhiro Nakazawa, "Broad-band spectral analysis of the Galactic Ridge X-ray Emission", Astrophysical Journal, vol. 753, 129, 2012

Takayuki Yuasa, Kazuhiro Nakazawa, Kazuo Makishima, Kei Saitou, Manabu Ishida, Ken Ebisawa, Hideyuki Mori, and Shin'ya Yamada, "White dwarf masses in intermediate polars observed with the Suzaku satellite", Astronomy & Astrophysics, vol. 520, A25, 2010

Supervisor's Foreword

Half a century of cosmic X-ray observations has been filled with enchanting mysteries. The Ph.D. thesis of Dr. Takayuki Yuasa deals with one of them: that the Milky Way is shining diffusely in X-rays as a narrow belt in the sky. Since its discovery in the 1970s, this phenomenon, called the Galactic Ridge X-ray Emission (hereafter GRXE), has attracted many investigators. In the 1980s, prominent X-ray emission lines from highly ionized iron atoms were detected in the GRXE spectra with the Japanese *Tenma* satellite. Therefore, the GRXE can be understood as arising from hot plasmas with a temperature of $\sim 10^8$ K. This has led to two alternative interpretations. One of them assumes that the GRXE is emitted by truly diffuse hot plasmas filling the interstellar space. The other considers the same phenomenon as a collection of a large number of discrete Galactic sources that are too faint and numerous to be individually resolved. However, the former idea presented a difficulty, because such plasmas would be too hot to be gravitationally bound to the Galaxy. The latter also involved a difficulty, namely, that such a population of hot plasma sources was not known. Thus, the interpretations have remained controversial.

Utilizing the unprecedented angular resolution of the *Chandra* X-ray Observatory of the United States, a considerable fraction of the GRXE surface brightness in a particular sky direction was recently resolved into discrete sources, for which mass-accreting magnetic white-dwarf (WD), binaries are suggested as the candidate population. In such a binary system, gas from the companion star falls onto two magnetic poles of the WD, forming therein X-ray emitting hot plasmas. On this basis, Dr. Yuasa decided to use the Japanese *Suzaku* mission to examine whether WD binaries can actually account for the spectrum and surface brightness of the GRXE. *Suzaku* is particularly suited for this purpose, because the X-ray Imaging Spectrometer onboard provides sensitive diagnostics of the iron line complex, while the co-aligned Hard X-ray Detector can measure the hard X-ray continuum of GRXE much better than ever.

Dr. Yuasa's thesis consists of two major parts. In the first part, he constructed a numerical model, which predicts X-ray spectra of mass-accreting magnetic WDs when the WD mass and the plasma metallicity are specified. The model assumes that hot plasmas, shock-created at a certain height above the WD poles, flow down along the magnetic fields onto the WD surface while cooling by emitting thermal bremsstrahlung continua and atomic emission lines. This model was successfully applied to 2–40 keV *Suzaku* spectra of about 20 magnetic WDs, which was observed mainly based on Dr. Yuasa's own proposals. The latter part of the thesis, indeed its highlight, is an attempt to apply the validated WD emission model to high-quality *Suzaku* spectra of the GRXE, accumulated over many sky fields near the Galactic center by carefully avoiding bright point X-ray sources. As a result, he has successfully demonstrated that the GRXE spectra can be reproduced by his own WD emission model with plausible mass and abundance parameters, on condition that another softer plasma emission (presumably from active stars) is added.

In this way, Dr. Yuasa has demonstrated spectroscopically that the GRXE, in particular in the $\sim$5 to 40 keV range, can be understood principally as a superposition of a large number of mass-accreting WD binaries. He has also shown that these binaries are numerous enough to constitute the GRXE surface brightness when integrated along the line of sight. This is indeed a monumental result that has settled the long-lasting controversy of the GRXE. In short, the X-ray view of the Milky Way repeated its much older visible-light recognition; what appeared as a diffuse optical belt to the naked eye was later understood as an assembly of uncountable faint stars. Based on these achievements, Dr. Yuasa was awarded the Student Research Prize in March 2011 by the Graduate School of Science, The University of Tokyo.

January 2012 Kazuo Makishima

Acknowledgments

I would like to thank Prof. Kazuo Makishima for guiding and encouraging me throughout my five years at the graduate school of The University of Tokyo. Discussions with him led me to study the Galactic X-ray background emission and deepened my understanding of its origin. I greatly enjoyed being in his group, improving my skills in experimental physics and celestial data analyses.

I also thank Lecturer Kazuhiro Nakazawa for discussions on technical details of the *Suzaku* instruments. Dr. Shin'ya Yamada has been a wonderful colleague with his knowledge of black holes, and modern computing paradigms, and his spirit of collaboration. His continuous encouragement helped me to pursue my research activities in experiments and X-ray astrophysics. I have been very happy to work with him in the same office for the past five years.

The significant efforts of past members of the High Energy Astrophysics Laboratory of The University of Tokyo, which were offered for developing the Hard X-ray Detector (*ASTRO-E*) and the Hard X-ray Detector II (*ASTRO-E2/ Suzaku*), made possible the present wide-band spectral study of the Galactic background emission with the highest quality ever. I acknowledge the satellite operation staff working in ISAS/JAXA and JAXA's Uchinoura Space Center in Kagoshima, Japan, for providing scientific data that were essential in my analysis.

Finally, I am grateful to my family for their constant support in many ways over the years.

Acknowledgments

Contents

Abbreviation

APEC	Astro Physical Emission Code
BGO	Bismuth Germanate ($Bi_4Ge_3O_{12}$)
CIE	Collisional Ionization Equilibrium
CV	Cataclysmic Variable
CXB	Cosmic X-ray Background
GRXE	Galactic Ridge X-ray Emission
GSO	Ce Doped Gadolinium Silicate (Gd_2SiO_5)
HXD	Hard X-ray Detector
IBIS	Imager on-Board the Integral Satellite
Integral	International Gamma-ray Laboratory
IP	Intermediate Polar
NXB	Non X-ray Background
PIN	*p*-Intrinsic-*n* Si Sensor
PSR	Post Shock Region
WD	White Dwarf
XIS	X-ray Imaging Spectrometer
XRT	X-Ray Telescope

Chapter 1
Introduction

1.1 Overview

In astrophysics, several types of apparently diffuse celestial "background" emissions have been detected from wide area of the sky in ranges of electromagnetic wavelengths. Table 1.1 summarizes them ordered by their typical wavelengths. As can be recognized from their names, i.e., Galactic or cosmic, some of them have origins in the Milky Way Galaxy, and the others come almost isotropically from extragalactic universe.

As exemplified by the contribution of the Cosmic Microwave Background studies to the modern cosmology (e.g., Spergel et al. 2003), a celestial background emission often provides essential information on structures of the Universe and the Galaxy. In particular, background photons with high energies (i.e., X-rays and gamma-rays) are expected to have deep connections with, for example, cosmic-ray acceleration and general relativity. Therefore, studying the background emission phenomena in high energies can potentially give results relevant to fundamental physics. A nice example of this is a fact revealed by a study of the Galactic diffuse Gamma-ray Emission with the *Fermi* gamma-ray observatory, that the spectra of cosmic-ray particles at an interstellar region 1 kpc away are not largely different from those directly observed at the Earth (Abdo et al. 2009).

The origins of most of the background emissions have been identified through observational and theoretical studies for years, and those of a few of them are still unclear. Among them, we concentrate, in the present thesis, on the Galactic X-ray Background Emission, or historically called the Galactic Ridge X-ray Emission (GRXE), of which the origin has been a mystery for more than 40 years since its discovery (e.g., Cooke et al. 1969). Although many attempts were made to investigate the GRXE, with its apparently diffuse nature, in terms of some truly diffuse phenomena in the interstellar space, recent X-ray observations with the highest angular resolution (e.g., Revnivtsev et al. 2009) are reinforcing an alternative view that the GRXE is mostly composed of numerous discrete point sources. Then, this idea must be examined not only via the X-ray imagery but also via X-ray spectroscopy, by

T. Yuasa, Suzaku *Studies of White Dwarf Stars and the Galactic X-ray Background Emission*, Springer Theses, DOI: 10.1007/978-4-431-54219-3_1, © Springer Japan 2013

Table 1.1 Background emission observed in various electromagnetic wavelengths

Name	Range[a]	Origin(s)
Radio background	0.1–10 m	Quasars, star-forming galaxies
Cosmic microwave background	0.01–1 cm	Big bang
(Galactic) Infrared foreground	several × 10–1000 μ	Stars and dusts in the Galaxy
Infrared background	0.1–1000 μ	Galaxies
Optical/ultraviolet background	1–10 eV	Dominated by the Zodiacal light (solar photons scattered by dust)
Galactic X-ray background	2–60 keV	Mostly discrete faint X-ray sources in the Galaxy
Cosmic X-ray background	2–200 keV	Active galactic nuclei (mostly)
Galactic gamma-ray emission	0.1–100 GeV	Interaction of cosmic-rays and interstellar gas + electromagnetic field
Cosmic gamma-ray background	0.1–100 GeV	Blazers (<50 %), star forming galaxies (?), and cosmological shocks (?)

[a]Typical wavelength/energy ranges in which the background emission is observed

synthesizing the GRXE spectrum using those of known X-ray star populations in the Galaxy. Aiming to make them clear, we perform detailed spectroscopy of the GRXE using new, high quality data taken with the *Suzaku* X-ray observatory.

When trying to spectroscopically decompose the GRXE, it becomes necessary to understand spectral properties of possible candidates for the discrete X-ray sources. Among them, magnetic accreting white dwarf (WD) stars outstand especially in hard X-rays, i.e., energies or frequencies above 10 keV or $\sim 2 \times 10^{18}$ Hz, respectively. Being one of the major endpoints of stellar evolution, a WD is a compact star in which gravity is counter-balanced by degenerate electron pressure, a manifestation of quantum statistical physics. The magnetic accreting WDs form an important subset of WDs. Since the GRXE spectrum measured by *Suzaku* covers the energies where the GRXE is substantially contributed by this type of stars, we first characterize individual spectra of magnetic accreting WDs located close to the Sun. This study not only provides important clues used in the following GRXE study, but also allows us to estimate physically intriguing quantities such as the mass of a WD residing in this kind of X-ray stars.

By referring to the result of the individual sources, we then analyze wide-band GRXE spectra, and show that they can be decomposed into contributions from several types of known, faint, but numerous X-ray sources. Based on this study, we also estimate the mean WD mass and the number density of magnetic accreting WDs in the Galaxy.

1.2 Outline

The thesis is organized as follows. In Chap. 2, we review the Galactic Ridge X-ray Emission (GRXE) and X-ray emitting binary stars containing WDs, and explain our approach towards the origin of the GRXE. To quantify X-ray spectra of magnetic accreting WDs as a function of the WD mass, we construct, in Chap. 3, a spectral model for accreting WDs by numerically solving hydrostatic shock relations. Chapter 4 is devoted to explanation of X-ray instrumentation used to observe spectra of the GRXE and magnetic accreting WDs. Chapter 5 describes WD masses derived by applying the constructed model to the *Suzaku* data. Chapters 3 and 5 are partially based on a paper already published as Yuasa et al. (2010), but the present work contains essential updates concerning the plasma cooling function. In Chap. 6, we analyze the GRXE spectrum obtained with deep *Suzaku* observations, and decompose it into several spectral components which have physical correspondents. Chapter 7 summarizes the result and discussions given in the previous chapters. Appendices follow the body text to show detailed spectral analysis results (Appendix A), full observation log (Appendix B), and supplementary analysis result of dwarf nova stars (Appendix C).

1.3 The X-ray/Gamma-ray Instruments

The present study stands not only on the latest observational data by *Suzaku* (Chap. 4), but also on the accumulation of knowledge obtained by many other X-ray and gamma-ray instruments. Since their names are often referred to in the following text, we briefly summarize in Table 1.2 their names and energy coverages. The list is not complete, and focuses on instruments which are relevant to the present study.

1.4 Expressions

1.4.1 Notations

In the following chapters, notations listed in Table 1.3 are used to refer to physical quantities.

1.4.2 The Energy Flux Unit "Crab"

In addition, the so-called "Crab unit" frequently appears in the following part of the thesis to express an energy flux from a celestial source in a certain energy range. The Crab nebula is a pulsar wind nebula in the Milky Way Galaxy, and is one of the brightest celestial X-ray source in the sky. Since its energy flux has been almost stable in the X-ray band (0.1–10 keV), many X-ray instruments used this as the standard candle for calibrating their effective area, instrumental response, and so on. Therefore, an energy flux of an X-ray source are sometimes normalized with that

Table 1.2 X-ray and gamma-ray instruments relevant to the present thesis

Satellite	Period	Instrument name	Energy coverage
Tenma	1983–1985	Gas Scintillator Proportional Counter (GSPC)	2–60 keV
Ginga	1987–1991	Large Area Counter (LAC)	3–20 keV
ASCA	1993–2001	Solid-state Imaging Spectrometer (SIS)	0.4–12 keV
		Gas Imaging Spectrometer (GIS)	0.8–12 keV
RXTE	1995–	Proportional Counter Array (PCA)	2–60 keV
		High Energy X-ray Timing Experiment (HEXTE)	15–250 keV
Chandra	1999–	High Resolution Mirror Assembly (HRMA)	
		+ AXAF CCD Imaging Spectrometer (ACIS)	0.2–10 keV
XMM-Newton	1999–	X-ray mirror	
		+ European Photon Imaging Camera (EPIC)	0.1–15 keV
INTEGRAL	2002–	Imager on-Board the *INTEGRAL* Satellite (IBIS)	15 keV–10 MeV
Swift	2004–	Burst Alert Telescope (BAT)	15–100 keV
Suzaku	2005–	X-Ray Telescope (XRT)	
		+ X-ray Imaging Spectrometer (XIS)	0.1–12 keV
		Hard X-ray Detector (HXD)	12–600 keV

Table 1.3 Notations for physical quantities followed by units

Variable	Unit	Notes
c	cm s^{-1}	Light speed
l, b	degree	Galactic longitude and latitude
$M_{\rm WD}$	g	Mass of a white dwarf star
$R_{\rm WD}$	cm	Radius of a white dwarf star
$M_{\odot}$, $M_{\rm sun}$	g	Mass of the Sun
$R_{\odot}$, $R_{\rm sun}$	cm	Radius of the Sun
z	cm	Height of an accretion column or a post-shock region
$z_{\rm s}$	cm	Height of a standing shock
ρ	g cm^{-3}	Plasma density
Λ	erg cm^{3} s^{-1}	Plasma cooling rate at a certain temperature.
k	eV K^{-1}	The Boltzmann constant
T	K	Plasma temperature
$T_{\rm s}$	K	Plasma temperature just behind the standing shock
a	g cm^{-2} s^{-1}	Specific mass accretion rate at the shock height
v	cm s^{-1}	Speed of bulk motion of an accreting plasma
f	1	Ratio of an accretion region over the surface area of a white dwarf star
F	erg cm^{-2} s^{-1}	Energy flux

of the Crab nebula to conveniently show how it is bright (or faint) compared to the standard candle.

For reference, we present energy fluxes of the Crab nebula. We assumed a spectral model consisting of a power-law function in which its differential photon flux f at an energy E is expressed as

$$f(E) = 11.2 \left(\frac{E}{1\ \mathrm{keV}} \right)^{-2.1} \quad \text{photons cm}^{-2}\ \text{s}^{-1}\ \text{keV}^{-1}.$$

An energy flux F can be obtained by integrating $f(E)E$ over a certain energy range as

$$F = \int_{E_{\mathrm{low}}}^{E_{\mathrm{max}}} f(E)E\mathrm{d}E.$$

Thus calculated energy fluxes are 2.5×10^{-8} erg cm^{-2} s^{-1} (in the energy band 2–10 keV), 4.6×10^{-8} erg cm^{-2} s^{-1} (2–50 keV), and 1.9×10^{-8} erg cm^{-2} s^{-1} (12–50 keV).

References

Abdo, A. A., Ackermann, M., Ajello, M., et al. 2009, ApJ, 703, 1249
Cooke, B. A., Griffiths, R. E., & Pounds, K. A. 1969, Nature, 224, 134
Revnivtsev, M., Sazonov, S., Churazov, E., et al. 2009, Nature, 458, 1142
Spergel, D. N., Verde, L., Peiris, H. V., et al. 2003, ApJS, 148, 175
Yuasa, T., Nakazawa, K., Makishima, K., et al. 2010, A&A, 520, A25

Chapter 2
Review of the Galactic X-ray Background Emission and White Dwarf Binaries

In this chapter, we review two major scientific subjects of the present study, the Galactic X-ray background emission or so-called Galactic Ridge X-ray Emission (GRXE), and X-ray binary systems harboring white dwarf (WD) stars. The latter is thought to have a close connection with the GRXE origin especially in hard X-ray wavelengths, and hence plays a crucial role in physically interpreting the X-ray spectrum of the GRXE.

2.1 The Galactic Ridge X-ray Emission

2.1.1 Discovery and Observational Characteristics

Since very early days of the X-ray astrophysics, two types of background emissions have been known to exist in the X-ray wavelength from $\sim$2 keV to several tens of keV. One of them shows isotropic surface brightness over the sky with little dependence on the Galactic coordinates. To stress this characteristic, this component is called the cosmic X-ray background (CXB). The CXB has been observationally revealed to be composed of plenty of active galactic nuclei (AGN) located in distant galaxies (e.g., Schmidt et al. 1998). Many studies have been carried out over 3 decades to resolve the CXB into discrete AGNs (e.g., Giacconi et al. 1979; Ueda et al. 1999; Mushotzky et al. 2000), and eventually, the resolved fraction has increased to more than 80 % (Moretti et al. 2003) thanks to the splendid angular resolution ($\lesssim 1''$) of the *Chandra* satellite (Weisskopf et al. 2002).

Contrarily to the CXB, the other background emission is only detected in the Galactic plane and the Galactic bulge regions (Cooke et al. 1969, 1970; Worrall et al. 1982; Warwick et al. 1985; Koyama et al. 1986; Yamauchi and Koyama 1993), and generally referred to as the Galactic Ridge X-ray Emission.

T. Yuasa, Suzaku *Studies of White Dwarf Stars and the Galactic X-ray Background Emission*, Springer Theses, DOI: 10.1007/978-4-431-54219-3_2, © Springer Japan 2013

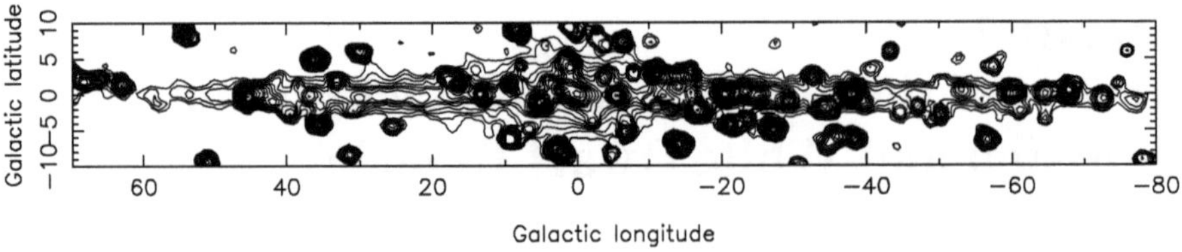

Fig. 2.1 An X-ray intensity map of the Galactic plane region measured by the *RXTE*/PCA in the energy band 3–20 keV. Contours start from an intensity of 10^{-11} erg cm^{-2} s^{-1} deg^{-2}, and logarithmically spaced with a factor of 1.4. Unresolved emission, which we call the GRXE, is clearly seen along the plane together with a number of bright point sources indicated by concentrations of the contours. Credit: Revnivtsev et al., A&A, Vol. 452, 169–178, 2006, reproduced with permission ©ESO

The GRXE apparently extends over several tens of degrees along the Galactic plane, and a few degrees in the Galactic latitude as can be seen in an X-ray intensity map presented in Fig. 2.1. The total luminosity and the surface brightness of the GRXE are $\sim 10^{38}$ erg s^{-1} (Koyama et al. 1986; Yamauchi and Koyama 1993) and $\lesssim$ several $\times\, 10^{-11}$ erg cm^{-2} s^{-1} deg^{-2} (Revnivtsev et al. 2006), respectively, in the energy band of 2–10 keV. The GRXE surface brightness is similar to or somewhat higher than that of the CXB, $\sim 2\times 10^{-11}$ erg cm^{-2} s^{-1} deg^{-2}, depending on regions; basically the GRXE is brighter near the Galactic center, and gets dimmer in off-center regions. Revnivtsev et al. (2006) revealed that the GRXE surface brightness measured by the *RXTE* satellite well correlates with that of the near infrared diffuse emission detected by *COBE*/DIRBE (Dwek et al. 1995) in the 1.25–4.9 μm range, suggesting close connection of the GRXE to late-type star population.

Using the *Tenma* satellite, Koyama et al. (1986) detected intense Fe Kα emission line (6.7 Kev) in the GRXE as presented in Fig. 2.2. Later, a strong enhancement in the Fe-K line surface brightness in the Galactic center region was discovered with the *Ginga* satellite (Koyama et al. 1989; Yamauchi and Koyama 1993). In addition to the Fe Kα, a number of emission lines from ionized lighter elements, such as Si, S, and Ar, have been clearly detected by several high energy resolution X-ray cameras (Kaneda et al. 1997; Muno et al. 2004). Because non-thermal emission processes, synchrotron or inverse-Compton scattering, do not produce atomic emission lines, these discoveries strongly suggest existence of optically-thin thermal plasma with a temperature of several keV (equivalently several $\times\, 10^7$ K) in the source of the GRXE. Emission lines from neutral metals have also been detected in the GRXE spectrum (Koyama et al. 1996; Nobukawa et al. 2009), especially in the Galactic center where the surface brightness is high. Notably, owing to the low background and large effective area of the *Suzaku* satellite, Ebisawa et al. (2008) discovered Kα line from neutral (or low ionized) Fe also in the GRXE of the Galactic plane. Since gas emitting this neutral F Kα line should be cool so that Fe is not ionized, these findings provided a challenging condition that X-ray-emitting hot plasma and cool gas coexist in the source of the GRXE.

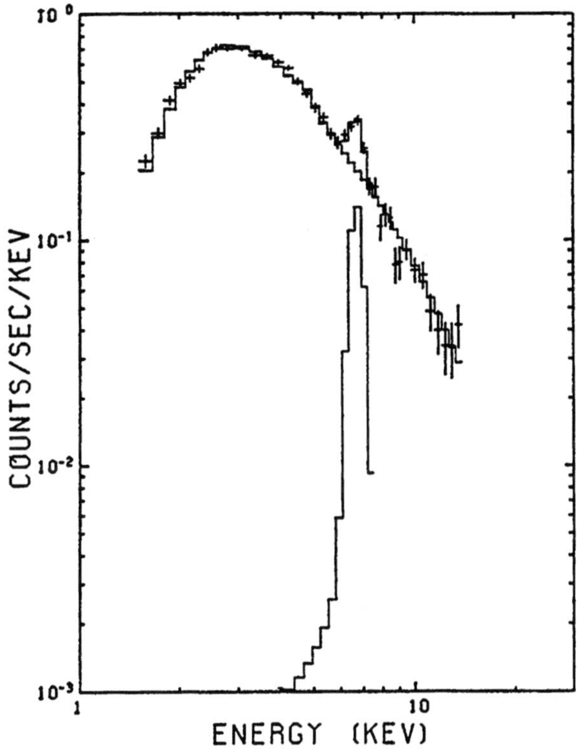

Fig. 2.2 A composite GRXE spectrum obtained by stacking eight observations on the Galactic plane ($280° < l < 340°$, $-5° < b < 5°$) with the *Tenma* satellite. *Crosses* are observed data shown without removing the instrumental response. *Curves* represent the best-fit bremsstrahlung continuum and the Fe emission line at 6.7 keV. Credit: Koyama et al. (1986), reproduced with permission of the author and the publisher

2.1.2 Proposed GRXE Origins

The origin of the GRXE has been controversial for more than 40 years. Many scenarios which employ different emission mechanisms have been proposed to explain the GRXE. They can be divided into two major groups depending on the assumed nature of the source, namely "Point Source" and "Diffuse" scenarios.

The former assumes that a collection of faint but numerous discrete X-ray point sources in the Galaxy composes the GRXE, similar to the case of the CXB. In this scenario, each X-ray-emitting material is gravitationally bound to the point source (i.e., X-ray star), and the source should have an optically-thin thermal X-ray spectrum to account for the observed intense atomic emission lines. On the other hand, in the latter scenario, diffuse X-ray emitting materials are considered to fill the interstellar space over the observed scales of the GRXE, i.e., several tens of degrees by a few degrees around the Galactic center. Below, we describe these scenarios in detail, stressing their advantages and disadvantages.

2.1.3 The "Point Source" Scenario

Luminous point sources with intrinsic X-ray luminosities higher than 10^{33}–10^{35} erg s^{-1} [1] cannot be the origin of the unresolved GRXE (e.g., Worrall et al. 1982), since they can be individually resolved with typical X-ray instruments. Therefore, several types of bright X-ray point sources, such as binaries containing a neutron star or a black hole accompanied with a mass-donating normal star, are rejected.

Below 10^{33} erg s^{-1}, a few X-ray source categories are plausible candidates for the origin; low-mass accreting WDs, active binaries, and coronally active late-type stars. Accreting WDs are binary systems which contain a WD as a compact object and a low-mass late-type star like the Sun as a mass donor (see Sect. 2.2.3 for detailed explanation). An active binary is a pair of late-type stars which form a close binary system, and emit X-rays via magnetic activities in their coronae as they rotate around each other. As shown in Fig. 2.3, these types of X-ray sources occupy low-luminosity

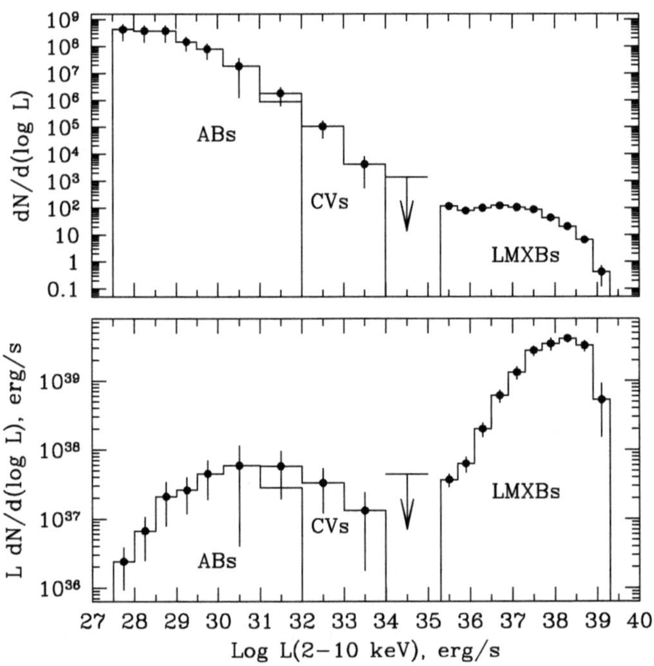

Fig. 2.3 *Upper panel* an X-ray luminosity function (i.e., number density against luminosity) of Galactic low-mass binaries, namely binaries involving low-mass stars. *Lower panel* a luminosity distribution of the binaries obtained by multiplying the luminosity L to the function in the upper panel. Credit: Sazonov et al., A&A, Vol. 450, 117–128, 2006, reproduced with permission ©ESO

[1] A distance of 8 kpc is assumed to calculate intrinsic fluxes. The distance corresponds to that to the Galactic center from the Sun.

regime among the Galactic X-ray source population, and hot plasmas bound to these sources, with typical temperatures of 10^7–10^8 K, emit X-rays via optically-thin thermal process, i.e., electron bremsstrahlung continuum and atomic emission lines. Worrall et al. (1982) estimated that the sum of accreting WDs and active binaries can account for ~15 % of the GRXE flux, in addition to a similar level of contribution from late-type stars, and attributed the remaining flux to discrete point sources with lower luminosities. Koyama et al. (1986) also discussed possible contribution from accreting WDs and active binaries based on their spectral similarity to the GRXE. Revnivtsev et al. (2006) pointed out that a wide-band GRXE spectrum can be well explained by the sum of individual spectra of these types of sources as presented in Fig. 2.4.

The only drawback of this scenario has been whether there are enough number of faint Galactic X-ray sources that can collectively explain the GRXE surface

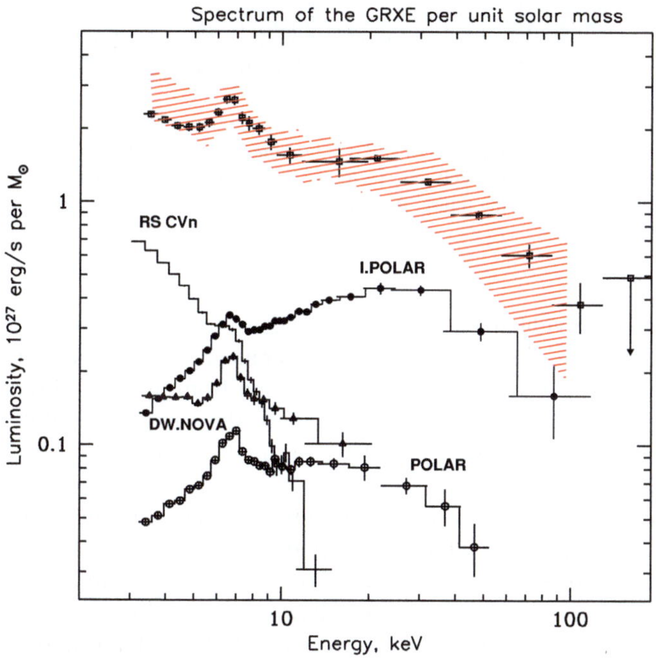

Fig. 2.4 A broad-band GRXE spectrum (*crosses with squares*) measured with the *RXTE*/PCA (3–20 keV) and the *INTEGRAL*/IBIS (20–100 keV). Ordinate indicates X-ray luminosity per stellar mass in the emission region. Predicted contributions from several types of low-luminosity X-ray point sources are also plotted by multiplying 0.5 for visibility. Labels correspond to X-ray source category names used in the present thesis; RS CVn = Active Binaries, DW. NOVA = Dwarf Novae, I.POLAR = Intermediate Polar. They are rescaled so that their relative contributions agree with the values derived from the X-ray luminosity function measured by Sazonov et al. (2006) as shown in Fig. 2.3. *Red* hatched region represents summed contribution from the point sources including uncertainties involved in the spectrum and the weighting factors. Credit: Revnivtsev et al., A&A, Vol. 452, 169–178, 2006, reproduced with permission ©ESO

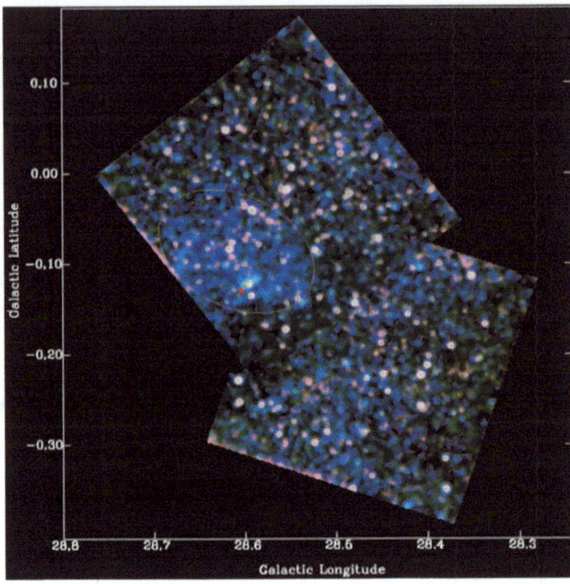

Fig. 2.5 An X-ray image of a Galactic ridge region obtained with 100 ks×2 *Chandra* observations. This pseudo color image is composed of *red*, *green*, *blue* images produced over the 0.5–2.0, 2.0–4.0, and 4.0–8.0 keV bands, respectively. In total, 274 discrete point sources were identified as marked with *small crosses* in the figure. Reproduced by permission of the American Astronomical Society

brightness and individually satisfy the luminosity limit ($<10^{33}$ erg s^{-1}). In previous studies which used instruments with moderate angular resolution (e.g., $>1'$), it was not possible to obtain a definitive conclusion from imaging observations (i.e., direct source counting). Like the CXB studies, the *Chandra* satellite provided a great leap also in the GRXE field thanks to its angular resolution in the X-ray wavelength ($<1''$).

Based on deep observations on the Galactic ridge using *Chandra*, Ebisawa et al. (2001) and Ebisawa et al. (2005) constructed a high resolution X-ray image of the GRXE as presented in Fig. 2.5, and counted point sources in the image. By comparing them with the detected total flux (unresolved emission plus resolved point source emission), they concluded that only ~10 % of the total GRXE flux have discrete point source origins under their detection limit (for point sources) of 3×10^{-15} erg cm^{-2} s^{-1} in the 2–10 keV band. This deficit was not solved even when they extrapolated source distribution further down to lower fluxes (n.b. see Revnivtsev and Sazonov 2007 for cautions on this extrapolation method).

Revnivtsev et al. (2007, 2009) claimed that more than 60 % (the Galactic center region) and 80 % (the Galactic bulge region) of the GRXE flux can be spatially resolved into discrete sources, and extrapolation of the source distribution can account for 100 %. Since the data set of Revnivtsev et al. (2009) consists of extremely-deep *Chandra* observations with a total exposure of 1 Ms, they achieved the ever highest

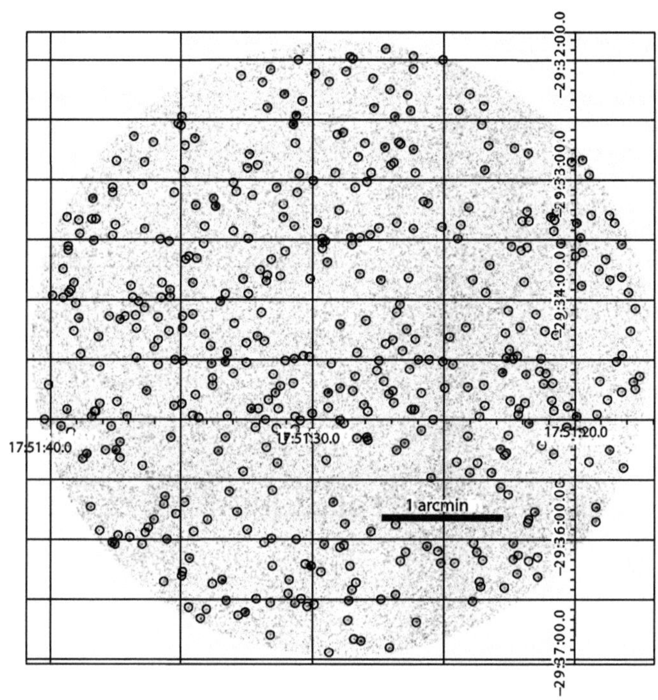

Fig. 2.6 An X-ray image of a Galactic ridge region obtained with the deep *Chandra* observation for 1 Ms. *Circles* mark 473 detected point sources. Reprinted by permission from Macmillan Publishers Ltd: Nature, Revnivtsev et al., copyright 2009

sensitivity to faint point sources, with a limiting flux of 10^{-16} erg cm^{-2} s^{-1} in the 0.5–7 keV band. Figure 2.6 presents an X-ray image of the Galactic ridge region analyzed by them. The limit is lower by more than an order of magnitude than that of Ebisawa et al. (2005). Therefore, based on their result, it is securely considered that at least 80 % of the total GRXE flux can be explained by discrete sources. If we concentrate on the prominent Fe Kα line structure in the 6.5–7.1 keV band, Revnivtsev et al. (2009) resolved as much as 88 ± 12 % of the detected flux into discrete point sources (see spectra shown in Fig. 2.7), leaving little room for truly diffuse emission to contribute to this line emission. Note that Revnivtsev et al. (2009) detected 473 sources in a circular region with a radius of $2.56'$ ($= 0.043°$). After subtracting a predicted contribution from extragalactic sources of 30–40, a surface source density is huge, $\sim$20 sources arcmin^{-2} ($\sim 7.2 \times 10^4$ sources deg^{-2}).

Some authors rejected contribution of active binaries and accreting WDs based on a consideration that intensities of the Fe emission line seen in their individual spectra are insufficient for explaining that of the GRXE spectrum (e.g., Yamauchi and Koyama 1993). However, they considered a single source population which could dominate the GRXE, instead of a more likely case of multiple source types constituting the GRXE (see Fig. 2.4 or e.g., Worrall et al. 1982). As described later

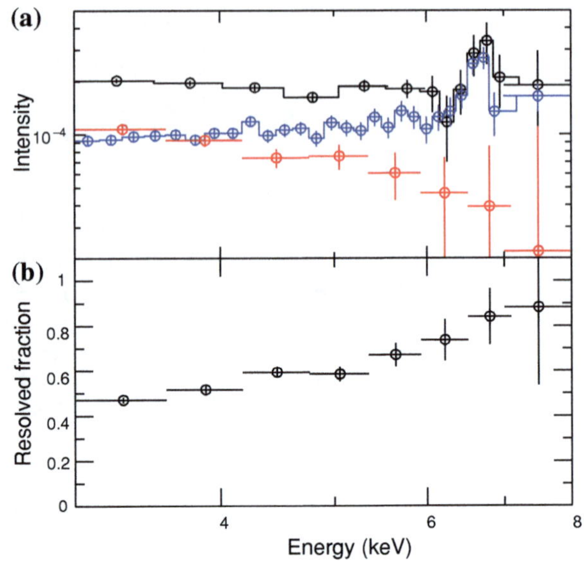

Fig. 2.7 **a** A GRXE spectrum extracted from the same data as the image (*black*). Extragalactic and instrumental background signals are subtracted. *Blue* and *red* crosses are composite spectrum of the detected point sources and residual against the total GRXE spectrum (i.e., unresolved component). **b** Fractions of the detected source contribution to the total GRXE signals. Reprinted by permission from Macmillan Publishers Ltd: Nature, Revnivtsev et al., copyright 2009

based on our analysis result, the deficit of the Fe emission line does not practically happen when interpreting a GRXE spectrum using a spectral model which considers both active binaries and accreting WDs (Sect. 6.2.3).

Apart from the imaging decomposition, Ebisawa et al. (2005) studied properties of faint point sources detected from their *Chandra* images. Spectra of the sources were constructed by accumulating photons of identified point sources. They also classified the sources into three groups, "soft", "medium", and "hard", according to the spectral hardness (i.e., color) defined as $(H - S)/(H + S)$ where S and H are photon counts in the soft (0.5–2.0 keV) and the hard (3–8 keV) bands, respectively. These groups had 98 (soft), 30 (medium), and 14 (hard) entries. In Fig. 2.8, we adopted the spectra thus created from their paper. The spectra of "hard" and "medium" sources clearly show the Fe emission line in the 6–7 keV band, and that of "soft" sources also gave a hint of that. The "soft" and "medium" spectra were fitted with two-temperature thermal plasma emission models, and gave the best-fit temperatures of $kT = 0.25 \pm 0.02$ and 2.2 ± 0.5 keV, and $kT = 0.23 \pm 0.10$ and 3.0 ± 0.4 keV, respectively. The higher temperature $kT = 2.2-3.4$ keV well corresponds to those of typical active binaries. The "hard" spectrum was interpreted as the sum of accreting WDs in quiescent states based on previous studies of Galactic faint point sources (e.g., Mukai and Shiokawa 1993), and nicely modeled using a power-law continuum and a Gaussian function for the Fe emission line both subjected to strong intrinsic absorption (Sect. 5.2.2).

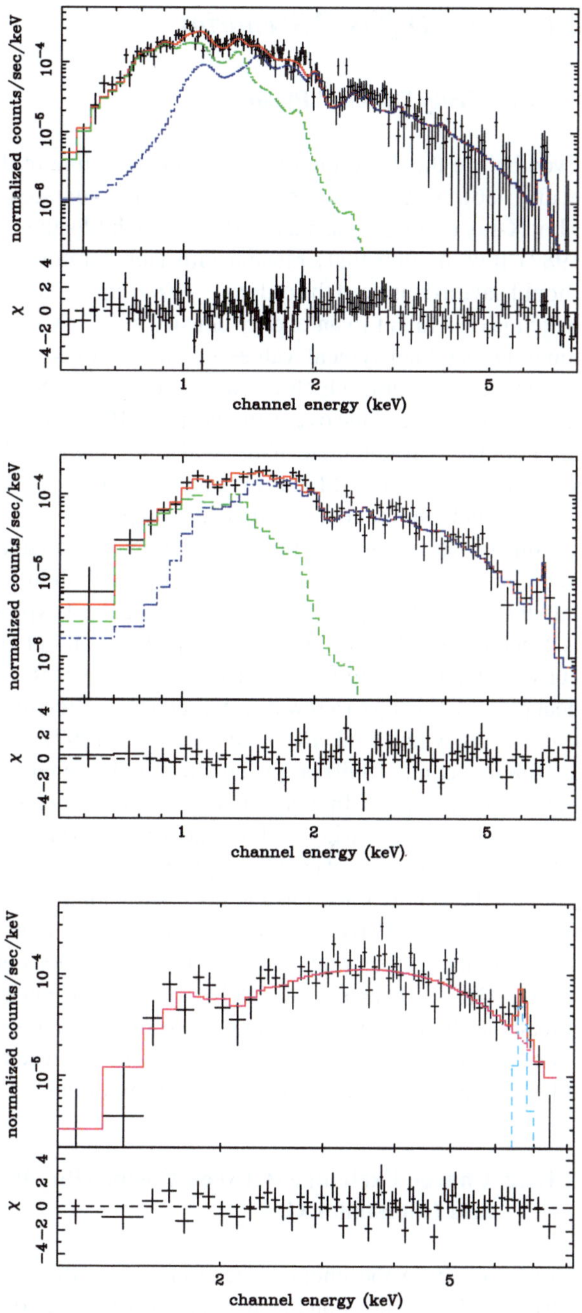

Fig. 2.8 Composite X-ray spectra of faint point sources detected in the deep Galactic ridge observations analyzed by Ebisawa et al. (2005). *Top*, *middle*, and *bottom* panels show stacked spectra of sources with "soft", "medium", and "hard" colors (see text). *Solid* and *dashed curves* are the best-fit model functions; two thermal models in the *top* and *middle* panels, and power-law continuum plus Gaussian in the *bottom*. For clarity, only those spectra of X-ray point sources which have near-infrared counterparts are used. Reproduced by permission of the American Astronomical Society

2.1.4 The "Diffuse" Scenarios

2.1.4.1 "Truly Diffuse" Plasma

Even before the detection of the Fe emission line in the GRXE spectrum (Koyama et al. 1986), many authors have been trying to explain the GRXE origin introducing a hot X-ray emitting plasma ($\sim 10^7$ K) which fills the interstellar space (Worrall et al. 1982). If we assume that a plasma with a temperature of 7 keV (Kaneda et al. 1997), it should have a density of the order of $\sim 10^{-3}$ cm^{-3}, and a pressure of $\sim 10^{-11}$ erg cm^{-3} which is equivalent to an energy density of ~ 5 eV cm^{-3}. This energy density is ~ 5 times higher than typical values of cosmic rays, interstellar magnetic fields, and interstellar medium, which are all ~ 1 eV cm^{-3}. Such a hot plasma would escape from the Galactic plane over a timescale of 10^4 years because its temperature exceeds the gravitational potential of the Galactic disk (~ 0.4 keV). Thus, "Truly Diffuse" scenario faces severe difficulties in energy injection (heating and replenishment of plasma) and confinement on the Galactic plane.

Supernova explosions are the most probable candidates which could inject energy to the putative hot plasma. However, even with their large amount of energy release into the interstellar space ($\sim 10^{51}$ erg), a required explosion rate to sustain the GRXE is 1 per several decades in the Galaxy (Yamasaki et al. 1997), and this exceeds our current understanding of ~ 1 per hundreds of years. In addition to this, supernovae heat plasma only up to a few keV, which is insufficient to explain the GRXE spectrum (>5 Kev; Koyama et al. 1986; Kaneda et al. 1997).

Tanuma et al. (1999) proposed a different scenario based on magnetic activity in the interstellar space. In their model, a plasma is heated via magnetic reconnection in the interstellar space triggered by a nearby supernova, and resulting helical magnetic tube confines the heated (~ 7 keV) plasma to the Galactic plane. This simultaneously satisfies the energy injection and the confinement problems. The required magnetic-field intensity, $\sim 30\,\mu$G, is actually observed in some limited environment such as star forming regions (e.g., $\sim 1\,\mu$G; Lai et al. 2003) and supernova remnants (e.g., ~ 15–$200\,\mu$G; Tanaka et al. 2008), and can be made consistent with the average interstellar field strength of a few μG if the plasma is confined within thin magnetic tubes like in the solar corona. However, general presence of such amplified magnetic tubes is not necessarily supported by observations.

2.1.4.2 Charge Exchange Between Cosmic-Ray Particles and Interstellar Gas

To be free from the energy injection and confinement problems, several authors attributed the Fe emission line to emission processes other than optically-thin thermal emission from a hot plasma. For atomic emission lines to be emitted, a transition of an electron between atomic energy levels from upper to lower should occur. Valinia et al. (2000) introduced low-energy cosmic-ray electrons with $E \sim 50$–100 keV which ionize heavy metals (e.g., Fe) in the interstellar medium, and creates vacancies in their atomic energy levels. Subsequent electron cascades to the vacancy produces Fe emission lines. Since the bulk speed of the interstellar medium and the temperature are not too high to broaden the line width by the Doppler effect, emission lines are expected to be narrow in their framework. Bremsstrahlung between the low-energy electrons and the interstellar medium (mainly hydrogen) produces continuum X-ray photons, and by taking this into account, required plasma temperature is reduced to <3 Kev. However, this model clearly fails to explain the Fe XXVI Kα line (H-like Kα) at 6.9 keV which was evident in the GRXE spectrum measured with the *ASCA*/SIS (Tanaka 2002; Ebisawa et al. 2008).

Similarly, low-energy cosmic-ray ions with energies of several tens of keV, particularly Fe, were considered as the origin of the Fe emission lines by Tanaka (2002). In this scenario, emission lines are produced via charge exchange between the cosmic-ray ions and the interstellar hydrogens (or hydrogen molecules). Electrons are transferred to the ion from the hydrogen in a collision, and then cascade to lower energy levels of the ion producing emission lines. The energies of line photons thus emitted should have broad distribution because of the bulk speed of the ions, which should amount to $\sim$5,000 km s^{-1}. Although such a "broadening" was reported to noticeable in an *ASCA* spectrum analyzed by him, later observations with recent instruments, for example by Ebisawa et al. (2008), revealed that the emission lines are narrow. Therefore, it is currently accepted that the low-energy cosmic-ray ions do not play a primary role in the GRXE production. In Sect. 6.2.2, we reconfirm non-existence of "broadened" Fe emission lines based on our *Suzaku* data. Critical constraints on the charge scenario are also found in Koyama et al. (2007) who measured center energies of the Fe lines precisely and discusses differences from values predicted from the charge exchange process; for example, the Fe XVV Kα emission line was centered at 6680 ± 1 eV which differs from 6666 ± 5 eV of the charge exchange case.

2.1.4.3 Non-Thermal Emissions from High Energy Particles

As reviewed in Worrall et al. (1982), electron synchrotron in the interstellar magnetic field and inverse-Compton scattering of microwave background or star lights on high energy electrons cannot effectively produce the observed GRXE flux, if realistic electron population or interstellar magnetic field strength are assumed. Furthermore, these non-thermal processes cannot account for the prominent Fe Kα emission lines.

Some authors (e.g.,Yamasaki et al. 1997; Valinia and Marshall 1998) argued that a hard X-ray spectrum of the GRXE, obtained with *Ginga* (3–16 keV) and *RXTE* (10–30 keV), smoothly connects to one obtained in the soft gamma-ray domain (50–500 keV using *CGRO*/OSSE). However it has been revealed that the CGRO data used in Valinia and Marshall (1998) were significantly contaminated by a few hard X-ray point sources, including the active galactic nucleus IGR J21247+5058 (Revnivtsev et al. 2006). Similarly, it is now clear that the observations of Yamasaki et al. (1997) also included numerous hard X-ray sources (especially below 100 keV) according to the point source catalog constructed from INTEGRAL/IBIS images (Krivonos et al. 2007b), and their GRXE spectral shape should be reexamined.

Regarding these later findings, the past-claimed smooth connection of their *GRXE* spectrum with their *diffuse* soft gamma-ray emission seems unrealistic from today's viewpoint. Wide-band spectral measurements with *INTEGRAL* have been showing that dominant components of the GRXE and unresolved gamma-ray emission change at 100 keV (Krivonos et al. 2007a; Türler et al. 2010) above 100 keV, gamma rays produced via interactions of cosmic-ray particles and the interstellar matter play an important role.

Masai et al. (2002) and Dogiel et al. (2002) discussed possibility of in-situ acceleration and X-ray emission from electrons accelerated, for example, in colliding molecular clouds. This scenario had attractive features that it only required a low temperature ($kT \sim 0.3$–0.4 keV) plasma, which can be bound to the Galactic disk by gravity, as an input for the acceleration, and energy injection needed to sustain such a plasma was thought to be affordable by supernovae. At this moment, there is no strong constraint or objection is proposed for this scenario, and neither is clear evidence supporting the sustained in-situ acceleration of electrons in the interstellar space. As explained later (Chap. 6), however, the GRXE spectrum strongly favors a plasma in the collisional ionization equilibrium which is totally different from the above condition (i.e., in-situ accelerated electrons).

Thus, the "Truly Diffuse" scenario still lacks a consistent explanation for the observed thermal spectrum, the energy injection, and the confinement of the GRXE. The recent findings of plenty of discrete sources in the deep *Chandra* images (Revnivtsev et al. 2009; Muno et al. 2009) severely limit contribution from possible "Truly Diffuse" emission to $<20\,\%$ (Revnivtsev et al. 2009).

2.1.5 The GRXE in the Galactic Center Region

The surface brightness of the GRXE rapidly increases toward the dynamical center of the Galaxy (e.g., Yamauchi and Koyama 1993) within Galactic longitudes of $\sim 2°$. As measured by Koyama et al. (1996), the GRXE spectral shape in this sky region is not largely different from those of the other regions (the Galactic bulge or the Galactic disk), but with stronger emission line intensities from Fe and other metals.

Extensive studies have been carried out in this particular region, the origin of this emission is still unclear, and it is unknown if this Galactic center X-ray emission and the GRXE have the same origin (Koyama et al. 2007; Nobukawa et al. 2010). Data analyzed in the present thesis do not contain this Galactic center emission, and therefore, we do not present any conclusive result relevant to this.

2.1.6 Our Approach Towards Understanding the GRXE

As explained above, the most probable explanation for the GRXE at present is the "Point Source" scenario which involves active binaries and accreting WDs, with possible contribution from X-ray emitting late-type stars like the Sun. However, it is not easy to further increase the point source sensitivity from what achieved in the deep *Chandra* observation by Revnivtsev et al. (2009). There is no proposed X-ray mission which has a better angular resolution than *Chandra*. Considering these situations, spatial decomposition (i.e., imaging) of the discrete faint point sources is almost done.

Questions still remain: How do individual types of X-ray sources compose the GRXE? and Are there any additional spectral components in its spectrum cannot be attributed to these types of sources?. To understand fractional contributions from several types of X-ray sources, we take an alternative approach, i.e., "detailed spectroscopy of the GRXE" in the present study. Essentially, the GRXE spectrum should be analyzed on a wide-band basis because some of the putative constituents, namely magnetic accreting WDs, are known to have characteristic spectral shapes in the hard X-ray energy band above 10 keV. High energy resolution is also necessary to understand the characteristic Fe emission lines which are thought to be emitted by several types of sources. At present, spectral decomposition of the GRXE, e.g., one shown in Fig. 2.7, is still primitive using spectra with an unresolved Fe emission line.[2] In pursuing this study, the *Suzaku* satellite (Chap. 4) provides an ideal mean, because of its unique capabilities such as large effective area with moderate energy resolution and stable/low particle background below 10 keV, together with the hard X-ray energy coverage up to 70 keV (Chap. 4).

When analyzing an GRXE spectrum obtained with *Suzaku*, the "Point Source" scenario can be a nice working hypothesis. Especially, in the hard X-ray energy band above 10 keV, magnetic accreting WDs, especially a subcategory of them called Intermediate Polars (Fig. 2.6), dominate the GRXE. Therefore, in Chap. 5, we study individual spectra of nearby Intermediate Polars in detail, and construct a spectral model applicable to them. Then, using the model and wide-band GRXE spectra covering the energy band 1–50 keV measured with *Suzaku*, we try to spectroscopically decompose the GRXE into contributions from several types of X-ray stars.

2.2 Accreting White Dwarf Stars (Cataclysmic Variables)

In this section, we review WD stars and binary systems containing them, because they are important contributor to the GRXE in the "Point Source" scenario.

2.2.1 White Dwarf Stars

A WD is a remnant of a star of which initial mass was in the range of $\sim$0.5–8 $M_{\odot}$.[3] In a core region of such a low or medium mass star, the hydrogen burning is followed by

---

[2] Note that they did not actually decompose the GRXE spectrum, but just compared a composite spectrum of several types of sources and the GRXE spectrum.

[3] Stars with masses below $\sim$0.5 $M_{\odot}$ have longer life times which are thought to as long as (or longer than) the age of the universe. Therefore, they are probably not the origin of WDs currently existing in the Galaxy.

helium fusion (triple-alpha process) which produces carbon and oxygen. The temperature ($\sim 10^7$–10^9 K) of the core, however, is not high enough for carbon and oxygen to further fuse into heavier elements. The carbon-oxygen core is surrounded by a helium-burning shell and a hydrogen-burning shell. After consuming all the helium in the core, there is no energy injection from nuclear reaction, and the core shrinks under the gravity, converting half the released gravitational energy into internal energy and radiating away the remaining half. Through this gravitational contraction, the classical gas pressure increases as $\propto n_e^{4/3}$ where n_e is the electron density in the stellar core, while the degenerate pressure of electrons as $\propto n_e^{5/3}$. At a certain point, the latter overwhelms the former, and prevents the star from shrinking further. Finally, the stellar core is left as a very dense star, namely a WD, while the outer envelope is lost into the interstellar space (often forming a "planetary nebula"). This makes the WD a physically important stellar object where microscopic degenerate pressure supports macroscopic gravitational force.

Based on surveys of nearby WDs, for example by Holberg et al. (2008) who collected objects within 13 pc from the Sun, a local space density of WDs is estimated to be $4.8 \pm 0.5 \times 10^{-3}$ pc^{-3}. This translates to a mass density of $3.2 \pm 0.3 \times 10^{-3}$ $M_\odot$ pc^{-3} if we assume a mean WD mass of 0.6 $M_\odot$. When compared to a mass density of non-degenerate stars ($\sim$0.03 $M_\odot$ pc^{-3}), the value accounts about $\sim$10 % of the total stellar mass density near the Sun (Holmberg and Flynn 2000). In the Galaxy, more than 9,000 WDs are detected (Eisenstein et al. 2006) from a sky region surveyed by the Sloan Digital Sky Survey (SDSS).

About 10 % of WDs have strong magnetic fields on the order of $B \sim 10^{4-9}$ G (Wickramasinghe and Ferrario 2005; Liebert et al. 2003), as manifested by the Zeeman splitting of hydrogen absorption lines seen in their optical/infrared spectra. These magnetic WDs posessing $B \gtrsim 5 \times 10^7$ G are thought to be descendents of magnetic Ap/Bp main-sequence stars (Schmidt et al. 2003; Ferrario and Wickramasinghe 2005). Progenitors of magnetic WDs with lower field strengths ($B < 10^4$ G) are not fully understood (Schmidt et al. 2003). The magnetism of WDs plays an important role especially in mass accreting WD binaries, because it alters the ways of mass transfer and accretion onto a WD (see below).

2.2.1.1 The Mass-Radius Relation of White Dwarf Stars

Roughly speaking, a WD is similar to the Earth in size and to the Sun in mass. Since it shines almost as a black body with an effective temperature of 10^4 – several $\times 10^4$ K, the total luminosity is proportional to its surface area. When compared with the Sun, a WD has two orders of magnitude smaller radii and hence four orders of magnitude lower luminosity, and this explains the reason why WDs are gathering at a $\sim$10 magnitudes lower (i.e., $\sim 10^{-4}$ dimmer) region in the Hertzsprung-Russell diagram shown in Fig. 2.9 compared to main-sequence (normal) stars like the Sun. Below, we review the WD mass-radius relation which is used later when constructing an X-ray spectral model of a WD that accretes mass from a companion star (Chap. 3).

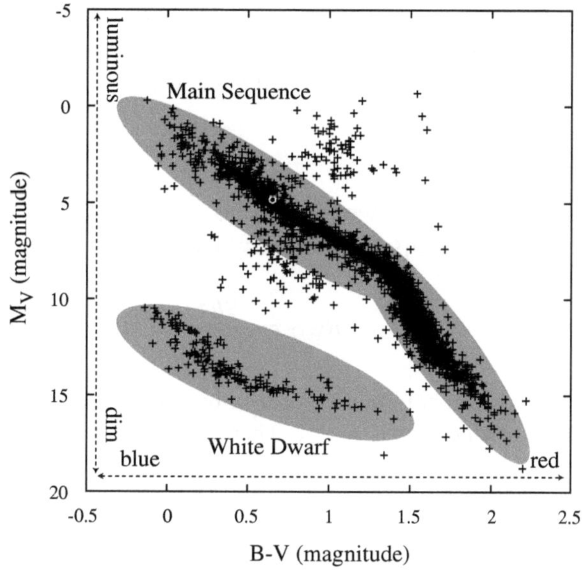

Fig. 2.9 The Hertzsprung-Russell diagram for stars near the Sun, produced using optical photometry data of Catalog of Nearby Stars version 3 (an update of Gliese and Jahreiß 1979). Abscissa and ordinate correspond to colors of stars (Blue minus Visible ≈ surface temperatures) and their intrinsic luminosities, respectively. The Sun is marked with a *white circle*

If we assume that a star is fully supported by the electron degenerate pressure, the mass and radius can be related as follows. According to a stellar structure model (e.g., one by Clayton 1968), the core pressure P_c of a WD with a mass of $M_{\rm WD}$ can be written using the core density ρ_c as

$$P_c \approx (\pi/36)^{1/3} G M_{\rm WD}^{2/3} \rho_c^{4/3}, \tag{2.1}$$

where G is the gravitational constant. This should be counter-balanced by the degenerate pressure of non-relativistic or relativistic electrons,

$$P_{\rm NR} = (h^2/5m_e)(3/8\pi)^{2/3} n_e^{5/3} \text{ (Non Relativistic)}, \tag{2.2}$$
$$P_{\rm Rel} = (hc/4)(3/8\pi)^{1/3} n_e^{4/3} \text{ (Relativistic)}, \tag{2.3}$$

depending on the energy regime. Here, h, m_e, and n_e are the Planck constant, mass and number density of electrons, respectively. By introducing mean molecular number of electrons μ, we can relate the electron density n_e and the mass density ρ using hydrogen mass $m_{\rm H}$ as

$$\rho = n_e m_{\rm H} \mu. \tag{2.4}$$

Assuming non-relativistic degenerate electrons, by equating Eqs. (2.1) and (2.2), and by substituting Eq. (2.4), we can reduce the core density as

$$\rho_c = \left(\frac{16\pi^3}{81}\right) \left(\frac{5m_e G}{h^2}\right)^3 (m_{\rm H}\mu)^5 M_{\rm WD}^2. \tag{2.5}$$

Employing a polytropic model for the stellar interior as $P(R) \propto \rho(R)^{5/3}$, where R and $\rho(R)$ are the distance from the star center and gas density, respectively, the mean density $<\rho>$ is related to the core density ρ_c as

$$<\rho>=\rho_c/6. \tag{2.6}$$

On the other hand, we can write as $<\rho>=3M_{\rm WD}/4\pi R_{\rm WD}^3$ where $R_{\rm WD}$ is the WD radius. Then, Eqs. (2.5) and (2.6) lead to

$$R_{\rm WD}=\frac{9h^2}{5m_{\rm e}G}(\pi^4 m_{\rm H}^5 \mu^5 2^5 M_{\rm WD})^{-1/3}. \tag{2.7}$$

As long as a helium core or a carbon-oxygen core is assumed, the mean molecular mass of electrons is $\mu = 2$. Therefore, numerical evaluation of this radius gives

$$R_{\rm WD}=9.5\times10^8\,{\rm cm}\left(\frac{M_{\rm WD}}{M_\odot}\right)^{-1/3} \tag{2.8}$$

$$=0.014R_\odot\left(\frac{M_{\rm WD}}{M_\odot}\right)^{-1/3}. \tag{2.9}$$

Based on this equation, a typical WD mass of 0.6 $M_\odot$ gives a WD radius of 1.6 % of that of the Sun, or 1.1×10^4 km.

For a WD mass of $M_{\rm WD} = 0.5\ M_\odot$, the Fermi energy of electron already exceeds several tens % of the electron rest mass. Therefore, the assumption of non-relativistic electrons, or the use of Eq. (2.3), may not be valid. After taking into account a relativistic-correction to the M-R relation, Nauenberg (1972) gave an analytical approximation

$$R_{\rm WD}=7.8\times10^8\left[\left(\frac{1.44M_\odot}{M_{\rm WD}}\right)^{2/3}-\left(\frac{M_{\rm WD}}{1.44M_\odot}\right)^{2/3}\right]^{1/2}\ {\rm cm}. \tag{2.10}$$

Figure 2.10 visualizes this relation.

As implied by Eq. (2.10), a WD with a mass higher than ~1.4 $M_\odot$ cannot be supported even by ultrarelativistic degenerate electrons. This maximum WD mass is called the Chandrasekhar limit, and computed more accurately on WDs with different chemical compositions (e.g., helium, carbon–oxygen, oxygen–neon–magnesium, and iron WDs) by Hamada and Salpeter (1961) and Panei et al. (2000) for example.

2.2.2 *White Dwarf Binaries*

Unlike the Sun which is an isolated main-sequence star, fairly large portion of stars in the Galaxy are thought to form binary systems. A star in a binary also evolve as

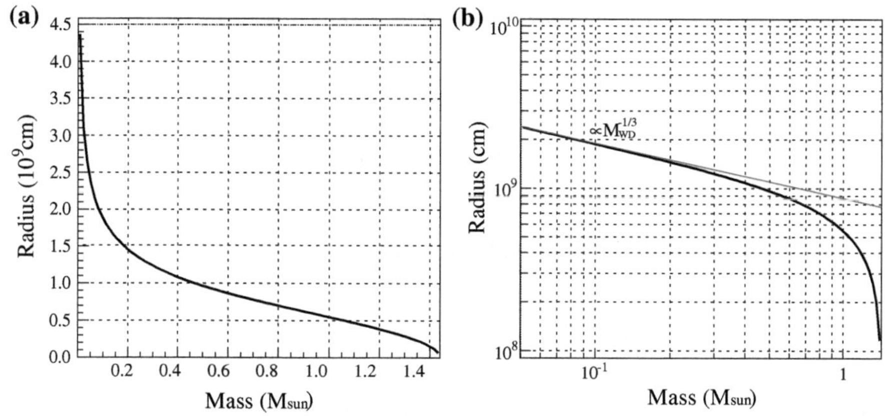

Fig. 2.10 **a** The WD mass-radius relation approximated by Nauenberg (1972). **b** The same as panel (**a**), but plotted in log–log axes. *Gray solid line* indicates a slope of $M_{WD}^{-1/3}$

isolated stars do, and if its mass is below 8 $M_\odot$, a WD is left as an end product of the binary. Actually, Sirius and Procyon are well-studied binary systems near the Sun, and each of them contains a WD as a companion to a main-sequence star.

2.2.2.1 Binary Kinematics

From Kepler's third law, an orbital separation of a binary system A and an orbital period P_{orb} can be related as

$$\left(\frac{2\pi}{P_{orb}}\right)^2 = \frac{G(M_1 + M_2)}{A^3}, \tag{2.11}$$

where M_1 and M_2 are the masses of component stars. Renormalizing with typical values, A can be expressed as

$$A = 3.5 \times 10^{10} \left(\frac{M_1 + M_2}{M_\odot}\right)^{1/3} \left(\frac{P_{orb}}{1\,\mathrm{h}}\right)^{2/3} \mathrm{cm}. \tag{2.12}$$

Some fraction of WD binaries are close-binary systems which have P_{orb} of several hours, and contains a low-mass main-sequence star as a companion to a WD. By substituting typical parameters such as a WD mass of $M_1 = M_{WD} = 0.6\ M_\odot$, a companion mass $M_2 = 0.3\ M_\odot$, and an orbital period $P_{ord} = 3$ hours, the above equation results in $A = 6.7 \times 10^{10}$ cm ($= 70\ R_{WD}$).

Using the inclination angle of the binary system i (angle measured between the pole of the orbital motion and our line of sight), we can write radial velocity amplitudes K_1 and K_2 of the stars (WD and its companion) as

$$K_1 = \frac{M_2}{M_1 + M_2}\left(\frac{2\pi}{P_{\rm orb}}\right) A \sin i, \tag{2.13}$$

$$K_2 = \frac{M_1}{M_1 + M_2}\left(\frac{2\pi}{P_{\rm orb}}\right) A \sin i. \tag{2.14}$$

By equating these, a frequently used binary parameter called "mass function"

$$f(M_1, M_2, i) = \frac{M_1^3 \sin^3 i}{(M_1 + M_2)^2} = \frac{P_{\rm orb} K_2}{2\pi G}, \tag{2.15}$$

can be obtained. Usually, $P_{\rm orb}$ and K_2 can be estimated from time variations of the optical brightness of the binary, and the Doppler shift of absorption lines which are thought to arise from the surface of the companion star, respectively. Substitution of these quantities into Eq. (2.15) gives important information on the mass ratio $q = M_1/M_2$.

2.2.2.2 Roche Potential

In a binary system, the gravitational attraction by the component stars and centrifugal force due to their orbital motion create a particular effective potential, called the Roche potential. As shown in Fig. 2.11, the potential has a saddle point L_1 between the two stars, and a separatrix passing through L_1. The two separate regions inside the separatrix are called the Roche lobes, and the gravity of each star works as a primary force within each lobe.

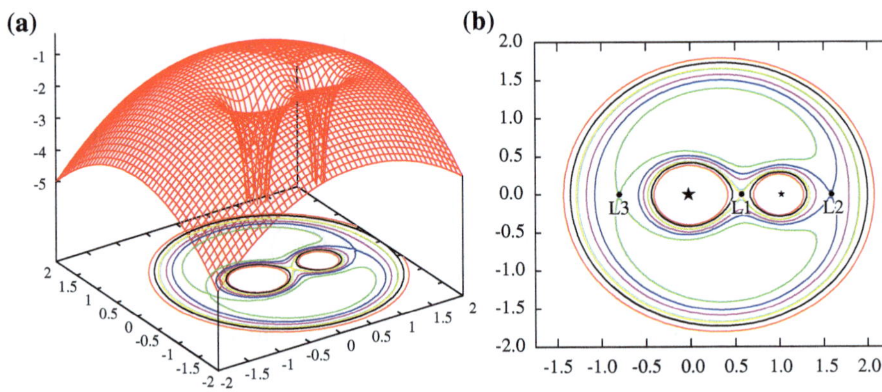

Fig. 2.11 **a** An example of the Roche potential calculated for a binary which contains two point-like stars with a mass ratio of 0.5; the star placed at the origin of the coordinate has twice a mass of the other which is located at (1, 0). **b** The same as panel **(a)**, but only shows the contour plot for clarity. *Black filled circles* indicate three Lagrange points, L1, L2, and L3, which are unstable saddle points of the potential

A typical size of the Roche lobe R_R can be estimated using an approximated formula of Eggleton (1983) as

$$R_R/A = \frac{0.49q^{2/3}}{0.6q^{2/3} + \ln(1 + q^{1/3})}, 0 < q < \infty. \tag{2.16}$$

For an assumed separation of $A = 0.1$ AU ($= 1.5 \times 10^7$ km), a mass ratio of $q = 2$ results in $R_R = 0.044$ AU ($= 6.6 \times 10^6$ km).

2.2.3 Cataclysmic Variables

If a star in a binary expands to fill its Roche lobe, or if the lobe contracts to the size smaller than a star inside it, gas in the outer layer of the star will be gravitationally unbound, and transferred to the other star via L1, or repelled to outward by the centrifugal force. These are called the Roche-lobe overflow, and the former situation is illustrated in Fig. 2.12. Especially the mass transfer and subsequent mass accretion onto the compact component (WD, neutron star, or black hole) are of great importance, because the liberated gravitational energy is huge, producing energetic radiations of high energy photons.

When a mass accretion from a normal star to a WD takes place in a binary, it is called "mass accreting WD", or historically, cataclysmic variable (CV) after its drastic optical variability. As drawn in Fig. 2.13, CVs are classified into two subcategories depending on magnetism of the WD; non-magnetic CVs and magnetic

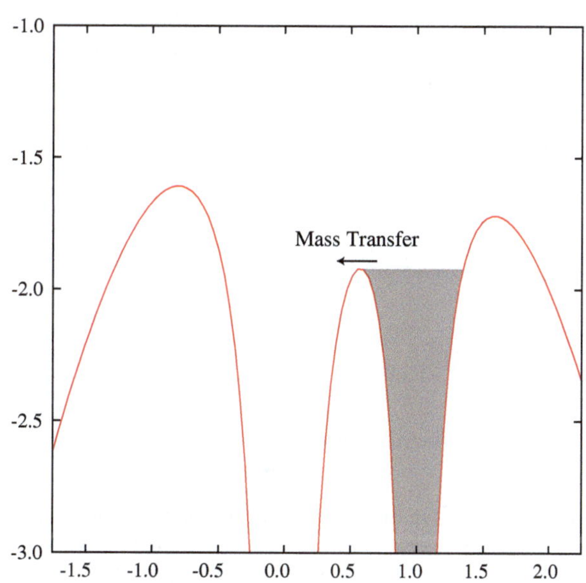

Fig. 2.12 The same Roche potential as shown in Fig. 2.11, sliced at $y = 0$. *Gray filled* region represents gas of a Roch-lobe-filling star which is gradually transferred to the companion via the inner Lagrange point L1

Fig. 2.13 Categories of CVs

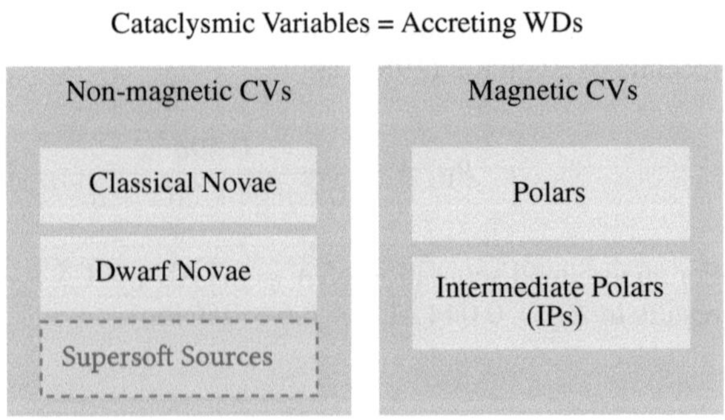

CVs. The former is further divided into classical novae, and dwarf novae based on the origin of their variability. The latter includes two subgroups, namely Polars and Intermediate Polars, which have different magnetic field strengths. Supersoft X-ray sources can be included in the non-magnetic CV category, although they exhibit stable (not cataclysmically variable) soft X-ray emission with energies of ~10–100 eV, and they are not main targets of the present study because of their scarcity in the Galaxy and special mass accretion mechanism (stable nuclear reaction on the WD surface).

2.2.3.1 Non-Magnetic CVs

Non-magnetic CVs, namely classical novae and dwarf novae, are close binary systems containing WDs with weak magnetic fields. These objects have been known from old ages because they change their luminosity drastically in visible light ($\sim 10^4$ times in classical novae and ~100 times in dwarf novae). For detailed reviews of classical novae and dwarf novae, Gallagher and Starrfield (1978) and Osaki (1996), respectively, may be reffered to.

Because the gas transferred from a companion star carries angular momentum, it forms an accretion disk around a WD as presented in Fig. 2.14. There, the angular

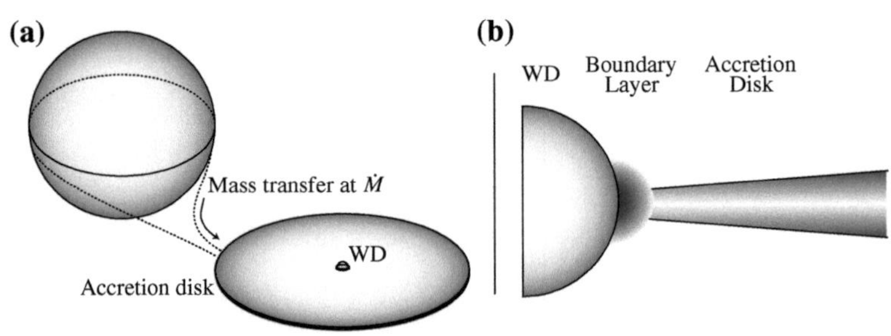

Fig. 2.14 a A typical geometrical configuration of non-magnetic CVs. b A close-up view of the boundary layer where inner edge of the accretion disk connects to the WD surface. This illustrates typical geometry of dwarf novae in which the boundary layer emits X-rays

momentum is somehow transported outwards, and the dynamical energy is dissipated via viscosity into thermal energy (i.e., heating the disk). Near the WD surface, Keplerian speed of the accreting gas becomes higher than that of the WD spin, and the gas should transfer its angular momentum further to accrete onto the WD. Although several mechanisms for losing the angular momentum in this transition region, or so-called boundary layer, have been proposed (Meyer and Meyer-Hofmeister 1994; Medvedev and Menou 2002), its physical configuration is not yet well understood.

In classical novae, the gas eventually settles on the WD surface, and gradually degenerates. After a long time of mass accumulation, the pressure and the temperature of the accreted matter become high enough to cause nuclear fusion, and there happens thermonuclear runaway, producing large amount of energy. Considerable amount of mass ($\sim 10^{-6}\ M_{\odot}$), including the accreted material and the surface layer of the WD, is expelled outward creating an expanding ejecta shell. This type of nova eruptions can be observed as dramatic intensity increase over wide ranges of electromagnetic frequencies from radio to X-rays, lasting for several tens to hundreds of days. Only several to ten classical novae happen in the Galaxy per year, and they cannot be a major contributor to the GRXE.

Dwarf novae steadily emit X-rays from the boundary layer in an optically quiescent state, and change their luminosities because of sudden increase of mass accretion rate, which in turn is caused by an instability in the accretion disk (Osaki 1996). Pandel et al. (2005) observed X-ray spectra of dwarf novae in quiescent states, and successfully reproduced the spectra in terms of thermal emission from a hot plasma with temperature gradient. To interpret the GRXE spectral components (Sect. 6.3.1), we also analyzed dwarf novae spectra observed with *Suzaku* in Appendix C. If we tentatively express their spectra with a single-temperature plasma model, typical plasma temperatures of 5–10 keV were obtained.

2.2.4 Magnetic CVs

Magnetic CVs are further classified into Polars and Intermediate Polars (IPs), according to their magnetic field strengths; 10^{7-9} G and 10^{5-7} G for the former and latter, respectively. Polars are so named after their strong cyclotron polarization in the optical and infrared wavelengths, and sometimes referred to as AM Her type following the prototype (for a review of Polars, see Cropper 1990). The magnetic field of a Polar is strong enough for the orbital and the spin periods of its WD to be synchronized ($P_{\rm orb} = P_{\rm spin}$). In IPs, the two periods are asynchronous, typically $P_{\rm spin} \sim 0.1\ P_{\rm orb}$ (for reviews of IPs, see Patterson 1994; Hellier 1996), and polarization is generally weak, although there are some recent reports of positive measurements (e.g., Butters et al. 2009).

In Fig. 2.15, we present the generally accepted binary and mass accretion geometry of magnetic CVs. It is known that accretion disks are present in most of IPs, while absent in Polars. This is explained in the following way. The accretion stream is affected by the magnetic field inside the Alfvén radius $R_{\rm A}$ where the magnetic

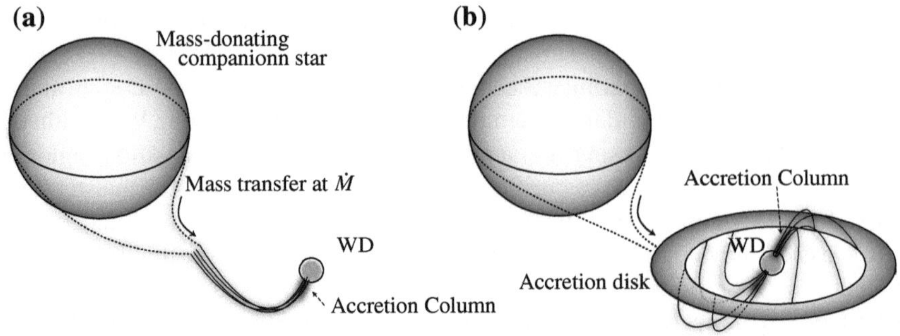

Fig. 2.15 Geometrical configurations of **a** Polar and **b** Intermediate Polar. WDs are enlarged in these figures compared to that in Fig. 2.14 for better visibility of the accretion columns

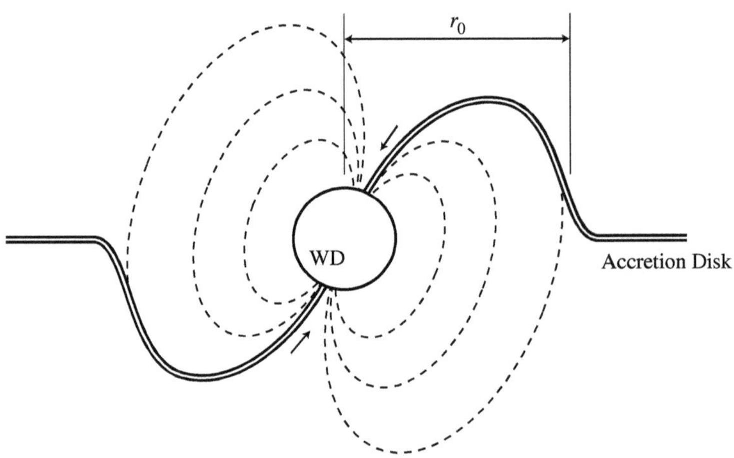

Fig. 2.16 A cross-sectional view of an accreting magnetic WD with a truncated accretion disk (i.e., Intermediate Polar)

pressure $P_B = B^2/8\pi$ equals the dynamical pressure of the accreting gas $P_{dyn} = \rho v^2$ (ρ: local gas density, v: local gas velocity). An accretion disk is totally disrupted inside a certain radius measured from the WD center $R_0 \sim R_A/2$ as schematically illustrated in Fig. 2.16.Since an accurate treatment of a possible inner radius R_0 is complicated, we just adopt a formula from literatures (Li et al. 1996; Ghosh and Lamb 1979),

$$R_0 = 0.52\mu_B^{4/7}(2GM_{WD})^{-1/7}\dot{M}^{-2/7}, \tag{2.17}$$

where $\mu_B = BR_{WD}$ is the magnetic moment of the WD, and $\dot{M}$ is the mass accretion rate (mass transferred from the companion in a unit time). Substitution of typical Polar values, such as $M_{WD} = 0.6\ M_\odot$, $B = 10^8$ G, and $\dot{M} = 5 \times 10^{-10}\ M_\odot$ (Sect. 5.3.3), to Eq. (2.17) results in $R_0 = 6.9 \times 10^5$ km ($\sim 80 R_{WD}$). This radius is larger than a typical orbital separation of Polars (Sect. 2.2.3), and therefore, an accretion disk is not formed. The mass transferred from the companion falls to the WD along the magnetic field line. Since R_0 is similar or less than the orbital separation in IPs (due to weaker magnetic field), most IPs have accretion disks, while the disk is truncated inside R_0. Unlike the disk, the accreting gas forms a column-like structure (hereafter accretion column) near the WD, and the gas almost freely falls onto the magnetic pole(s) of the WD channeled along its magnetic field.

The freely-falling gas has a speed $v \sim \sqrt{2GM_{\rm WD}/R}$; for instance, $M_{\rm WD} = 0.6\,M_\odot$ and $R = 1.1R_{\rm WD} = 9.5 \times 10^8$ cm give $v = 4.1 \times 10^8\,{\rm cm\,s^{-1}}$. Since this far exceeds the thermal speed of the gas near the WD surface, a stationary shock stands to convert substantial fraction of the kinematic energy of the bulk gas motion into thermal energy (Sect. 3.2). Across the shock, the gas is heated to a temperature of $kT > 10$ Kev ($T \gtrsim 1.2 \times 10^8$ K), and forms a post-shock region (PSR) with a temperature gradient, wherein the gas descends while it cools via the optically-thin thermal X-ray emission, i.e., continuum by electron bremsstrahlung and atomic emission lines from ionized ions (Aizu 1973; Frank et al. 1992; Wu 1994; Wu et al. 1994). As a result, the total spectrum from a PSR is observed as the sum of multi-temperature emission components (e.g., Ishida et al. 1994). For example, Fig. 2.17 shows an X-ray spectrum of the IP V1223 Sgr, measured by *RXTE* over the 3–100 keV band. The Fe emission line complex is evident at 6–7 keV, although its fine structure remains unresolved due to the instrumental energy resolution. The continuum which extends above energies of several tens of keV manifests a plasma with a very high temperature of $kT > 10$ Kev ($>10^8$ K) in the PSR. The position (i.e., energy) of the bending of the continuum represents the highest temperature, or the temperature just beneath the shock, in the PSR.

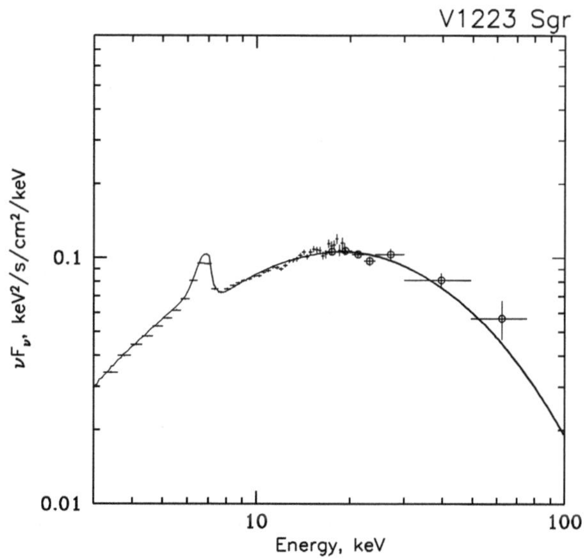

Fig. 2.17 A wide-band X-ray spectrum of the IP V1223 Sgr measured by *RXTE*, where the PCA and the HEXTE data are plotted as *black crosses* and *open circles*. The instrumental response is removed, and therefore, ordinate indicates to the spectral energy density per decade (or νF_ν, often used in astrophysics). Credit: Suleimanov et al., A&A, Vol. 435, 191–199, 2005, reproduced with permission ©ESO

Although magnetic CVs occupy a small fraction (~10–25 %) of all CVs, they dominate in the hard X-ray sky above $E > 10$ Kev, and most (~80 %) of CVs detected in recent hard X-ray surveys with *INTEGRAL*/IBIS and *Swift* are magnetized (Barlow et al. 2006; Landi et al. 2009; Bird etal. 2010), including both known (e.g., Revnivtsev et al. 2004) and new objects (Ajello et al. 2006; Bonnet-Bidaud et al. 2007; Butters et al. 2007; Evans et al. 2008; Sazonov et al. 2008; Landi et al. 2009).

According to recent X-ray studies (Sazonov et al. 2006; Revnivtsev et al. 2008), magnetic CVs form an important X-ray source population in the lower luminosity range ($L_X \sim 10^{31-34}$ erg s^{-1}; Fig. 2.3), while low-mass X-ray binaries with neutron stars and black hole binaries dominate in the high-luminosity range. As explained above, magnetic CVs, with their large numbers, are supposed to significantly contribute to the GRXE especially in the hard band, because of their spectral resemblance to that of the GRXE (Revnivtsev et al. 2006; Krivonos et al. 2007a) emitting hard continua and the Fe K lines.

2.2.5 Measuring the White Dwarf Mass

As presented in Fig. 2.10, heavier WDs have smaller radius, and therefore, deeper gravitational potential depths than lighter ones. The maximum temperature of a PSR of a magnetic CV is almost proportional to the star's gravitational potential depth, harder X-ray spectra (i.e., higher temperature emission) are expected from systems harboring heavier WDs. Thus, observed X-ray spectral shapes can be translated into the WD masses.

This is, in turn, very important for the GRXE study because the GRXE spectrum in the hard X-ray band (above $E > 10$ keV) is dominated by accreting magnetic WDs. If we obtain a reliable "averaged magnetic CV spectrum" in the hard X-ray band, we can directly compare it with the GRXE spectrum testing if the latter is reasonably described by the former, and the comparison will provide an immediate (in)validation of the "Point Source" scenario of the GRXE origin.

For enabling this, it is mandatory to collect a number of X-ray spectra of magnetic CVs, and to obtain comprehensive interpretation of the individual spectra before comparing them with that of the GRXE. The mass of a WD can be obtained as a by-product in such a study, which is also an important physical parameter related to the evolution of binaries and probably type Ia supernovae. Below, we review several mass estimation methods which are applicable to isolated WDs or binary systems containing a WD. In Chap. 3, an X-ray spectral model of IPs is constructed to realize the third method, and we apply it to actual data in Chap. 5.

2.2.5.1 Based on Binary Kinematics Measurements

A mass ratio q can be derived from Eq. (2.15) if the orbital period of the system $P_{\rm orb}$ and the radial velocity of the companion K_2 are measured. In CVs, we assume that the companion star fills its Roche lobe since there should be mass transfer to the WD via the Roche-lobe overflow. This provides another information on the size of the companion as $R_2 \approx R_R$, where R_R is the Roche lobe radius of the companion (not that of the WD as calculated in Sect. 2.2.3).

Independently, the mass-radius relation of main-sequence stars is empirically obtained as

$$\frac{R_2}{R_\odot} = 0.89 \left(\frac{M_2}{M_\odot}\right)^{0.89} \quad (\text{for } M_2 < 1.66\, M_\odot) \tag{2.18}$$

as reported by Demircan and Kahraman (1991) for example. Equating R_2 with R_R obtained in Eq. (2.16) followed by a substitution of q yields a function $M_2(P_{\rm orb})$. From the mass ratio q and M_2, $M_{\rm WD}$ can be estimated.

As plotted in Fig. 5.12, about 100 CVs have $M_{\rm WD}$ estimations based on this method (Ritter and Kolb 2003). However, the inclination angle i cannot be constrained well except for in eclipsing systems, and this still leaves rather large uncertainty in the $M_{\rm WD}$ estimation. When i is totally unknown, only the lower limit of $M_{\rm WD}$ can be obtained.

2.2.5.2 Based on Surface Gravity Measurements

The surface of a WD is covered with a hydrogen or helium layer of which the mass is $\lesssim 1\,\%$ of $M_{\rm WD}$, and they produces absorption lines in its optical/infrared spectrum. Since the surface gravity of a WD is fairly high ($\sim 3 \times 10^5$ times that of the Earth), the absorption lines are broadened due to atomic electron energy levels differently modified by the Stark effect (Bergeron et al. 1992).

Shapes of the broadened absorption lines can be predicted as a function of the surface gravity $g = GM_{\rm WD}/R_{\rm WD}^2$ and effective temperature $T_{\rm eff}$ if vertical profile of the density of the layer (i.e., an atmospheric model) is assumed (e.g., Bergeron et al. 1992; Kepler et al. 2007). Inversely, g and $T_{\rm eff}$ can be obtained by fitting the predicted absorption line shapes to an observed spectrum. To obtain $M_{\rm WD}$ and $R_{\rm WD}$ separately, the theoretical mass-radius relation such as Eq. (2.10) should be utilized. Among many studies which applied this method to actual data, Kepler et al. (2007) constructed the largest sample of 1,733 isolated WDs based on the SDSS data, and reported a mean WD mass of $M_{\rm WD} = 0.593 \pm 0.0016\, M_\odot$.

2.2.5.3 Based on Shock Temperature Measurements

Especially in a magnetic CV, X-ray spectroscopy provides an independent method since its X-ray spectrum is primarily determined by the shock temperature, which in turn reflects the gravitational potential, i.e., the $M_{\rm WD}/R_{\rm WD}$ ratio. Combining an X-ray-estimated $M_{\rm WD}/R_{\rm WD}$ with a theoretical WD mass-radius relation, such as Eq. (2.10), we can determine $M_{\rm WD}$.

The X-ray spectroscopic approach is in practice subdivided into two methods, both developed first in Japan; continuum method and emission line method. The former requires a sensitive hard X-ray detector since the relevant spectral feature, namely exponential cutoff which reflects the shock temperature, appears in the hard X-ray range (energies >10 keV; see Fig. 2.17). The latter relies on the resolved emission lines (e.g., from Fe), hence requiring a detector with a relatively high energy resolution such as an X-ray CCD camera. The continuum method has been applied to the data taken with *Ginga* (Ishida 1991; Cropper et al. 1998), *RXTE* (Ramsay 2000; Suleimanov et al. 2005, 2008), and *Swift* (Brunschweiger et al. 2009; Landi et al. 2009). Fujimoto and Ishida (1997) and Ezuka and Ishida (1999) applied the emission line method to the *ASCA* data, deriving metal abundances as well. Among these, Suleimanov et al. (2005) and Brunschweiger et al. (2009) are nice reference for the present study because they used data from modern hard X-ray detectors, and the numbers of systems they studied, 14 and 22 IPs, respectively, are large compared to others. Typical statistical fitting errors of these studies are 0.05–0.10 $M_{\odot}$, and the derived $M_{\rm WD}$ is distributed in the 0.45–1.10 $M_{\odot}$ range.

As already mentioned in Sect. 2.1.6, *Suzaku* has a high sensitivity over the wide energy band (0.2–600 keV) and a high energy resolution below 10 keV. These characteristics are unprecedented in previous X-ray observatories, and thus, we expect that *Suzaku* data provide significantly advanced information on the masses of magnetic CVs.

2.2.5.4 Other Methods

We briefly review several other methods which also allow us to estimate the mass of a WD but with some restrictions. Kawaler (1990) reported a method which utilizes information of pulsations of a WD. However, applicability of this method is limited to very hot objects, and therefore only few practical estimations have been made. For example, Winget et al. (1991) estimated $M_{\rm WD} = 0.586 \pm 0.003\ M_{\odot}$ for PG 1159–035, whereas the associated error does not contain systematic uncertainties which the method involves.

The gravitational redshift (of absorption line) also gives a mass-radius ratio of a WD, and this can be converted to $M_{\rm WD}$ if we further use the theoretical mass-radius relation. Since an accretion disk emission buries that from a WD surface, this method should be applied to an isolated object. In measuring the gravitational redshift, it is necessary to subtract the object's intrinsic radial velocity which is not always precisely available.

In a nearby WD ($\lesssim$100 pc) which has a trigonometric parallax measurement, the distance to the WD can be derived, and it converts an observed energy flux to the intrinsic luminosity. Using the Stefan-Boltzmann law, the surface area of the star and hence the radius are calculated from the luminosity, since we assume that a WD radiates as a black body. Combination of the radius and the theoretical mass-radius relation or an observational mass-radius ratio yields the mass of the WD. Vauclair et al. (1997) derived the masses of 20 WDs using the parallax measurements by *HIPPARCOS*, along with additional mass-radius ratios based on the surface gravity method. Typical errors associated with their mass estimates are 0.02–0.15 $M_{\odot}$.

References

Aizu, K. 1973, Progress of Theoretical Physics, 49, 1184
Ajello, M., Greiner, J., Rau, A., et al. 2006, The Astronomer's Telegram, 697, 1
Barlow, E. J., Knigge, C., Bird, A. J., et al. 2006, MNRAS, 372, 224
Bergeron, P., Saffer, R. A., & Liebert, J. 1992, ApJ, 394, 228
Bird, A. J., Bazzano, A., Bassani, L., et al. 2010, ApJS, 186, 1
Bonnet-Bidaud, J. M., de Martino, D., Falanga, M., Mouchet, M., & Masetti, N. 2007, A&A, 473, 185
Brunschweiger, J., Greiner, J., Ajello, M., & Osborne, J. 2009, A&A, 496, 121
Butters, O. W., Barlow, E. J., Norton, A. J., & Mukai, K. 2007, A&A, 475, L29
Butters, O. W., Katajainen, S., Norton, A. J., Lehto, H. J., & Piirola, V. 2009, A&A, 496, 891
Clayton, D. D. 1968, Principles of stellar evolution and nucleosynthesis, ed. Clayton, D. D.
Cooke, B. A., Griffiths, R. E., & Pounds, K. A. 1969, Nature, 224, 134
Cooke, B. A., Griffiths, R. E., & Pounds, K. A. 1970, in IAU Symposium, Vol. 37, Non-Solar X- and Gamma-Ray Astronomy, ed. L. Gratton, 280–+
Cropper, M. 1990, Space Science Reviews, 54, 195
Cropper, M., Ramsay, G., & Wu, K. 1998, MNRAS, 293, 222
Demircan, O. & Kahraman, G. 1991, Ap&SS, 181, 313
Dogiel, V. A., Inoue, H., Masai, K., Schönfelder, V., & Strong, A. W. 2002, ApJ, 581, 1061
Dwek, E., Arendt, R. G., Hauser, M. G., et al. 1995, ApJ, 445, 716
Ebisawa, K., Maeda, Y., Kaneda, H., & Yamauchi, S. 2001, Science, 293, 1633
Ebisawa, K., Tsujimoto, M., Paizis, A., et al. 2005, ApJ, 635, 214
Ebisawa, K., Yamauchi, S., Tanaka, Y., et al. 2008, PASJ, 60, 223
Eggleton, P. P. 1983, ApJ, 268, 368
Eisenstein, D. J., Liebert, J., Harris, H. C., et al. 2006, ApJS, 167, 40
Evans, P. A., Beardmore, A. P., & Osborne, J. P. 2008, The Astronomer's Telegram, 1669, 1
Ezuka, H. & Ishida, M. 1999, ApJS, 120, 277
Ferrario, L. & Wickramasinghe, D. T. 2005, MNRAS, 356, 615
Frank, J., King, A., & Raine, D. 1992, Accretion power in astrophysics., ed. K. A. R. D. Frank, J.
Fujimoto, R. & Ishida, M. 1997, ApJ, 474, 774
Gallagher, J. S. & Starrfield, S. 1978, ARA&A, 16, 171
Ghosh, P. & Lamb, F. K. 1979, ApJ, 232, 259
Giacconi, R., Bechtold, J., Branduardi, G., et al. 1979, ApJ, 234, L1
Gliese, W. & Jahreiß, H. 1979, A&AS, 38, 423
Hamada, T. & Salpeter, E. E. 1961, ApJ, 134, 683
Hellier, C. 1996, in Astrophysics and Space Science Library, Vol. 208, IAU Colloq. 158: Cataclysmic Variables and Related Objects, ed. A. Evans & J. H. Wood, 143
Holberg, J. B., Sion, E. M., Oswalt, T., et al. 2008, AJ, 135, 1225
Holmberg, J. & Flynn, C. 2000, MNRAS, 313, 209
Ishida, M. 1991, PhD thesis, The University of Tokyo.
Ishida, M., Mukai, K., & Osborne, J. P. 1994, PASJ, 46, L81
Kaneda, H., Makishima, K., Yamauchi, S., et al. 1997, ApJ, 491, 638
Kawaler, S. D. 1990, in Astronomical Society of the Pacific Conference Series, Vol. 11, Confrontation Between Stellar Pulsation and Evolution, ed. C. Cacciari & G. Clementini, 494–511
Kepler, S. O., Kleinman, S. J., Nitta, A., et al. 2007, MNRAS, 375, 1315
Koyama, K., Awaki, H., Kunieda, H., Takano, S., & Tawara, Y. 1989, Nature, 339, 603
Koyama, K., Hyodo, Y., Inui, T., et al. 2007, PASJ, 59, 245
Koyama, K., Maeda, Y., Sonobe, T., et al. 1996, PASJ, 48, 249
Koyama, K., Makishima, K., Tanaka, Y., & Tsunemi, H. 1986, PASJ, 38, 121
Krivonos, R., Revnivtsev, M., Churazov, E., et al. 2007a, A&A, 463, 957
Krivonos, R., Revnivtsev, M., Lutovinov, A., et al. 2007b, A&A, 475, 775
Lai, S., Girart, J. M., & Crutcher, R. M. 2003, ApJ, 598, 392

Landi, R., Bassani, L., Dean, A. J., et al. 2009, MNRAS, 392, 630
Li, J., Wickramasinghe, D. T., & Ruediger, G. 1996, ApJ, 469, 765
Liebert, J., Bergeron, P., & Holberg, J. B. 2003, AJ, 125, 348
Masai, K., Dogiel, V. A., Inoue, H., Schönfelder, V., & Strong, A. W. 2002, ApJ, 581, 1071
Medvedev, M. V. & Menou, K. 2002, ApJ, 565, L39
Meyer, F. & Meyer-Hofmeister, E. 1994, A&A, 288, 175
Moretti, A., Campana, S., Lazzati, D., & Tagliaferri, G. 2003, ApJ, 588, 696
Mukai, K. & Shiokawa, K. 1993, ApJ, 418, 863
Muno, M. P., Baganoff, F. K., Bautz, M. W., et al. 2004, ApJ, 613, 326
Muno, M. P., Bauer, F. E., Baganoff, F. K., et al. 2009, ApJS, 181, 110
Mushotzky, R. F., Cowie, L. L., Barger, A. J., & Arnaud, K. A. 2000, Nature, 404, 459
Nauenberg, M. 1972, ApJ, 175, 417
Nobukawa, M., Koyama, K., Matsumoto, H., & Tsuru, T. G. 2009, PASJ, 61, 93
Nobukawa, M., Koyama, K., Tsuru, T. G., Ryu, S. G., & Tatischeff, V. 2010, PASJ, 62, 423
Osaki, Y. 1996, PASP, 108, 39
Pandel, D., Córdova, F. A., Mason, K. O., & Priedhorsky, W. C. 2005, ApJ, 626, 396
Panei, J. A., Althaus, L. G., & Benvenuto, O. G. 2000, A&A, 353, 970
Patterson, J. 1994, PASP, 106, 209
Ramsay, G. 2000, MNRAS, 314, 403
Revnivtsev, M., Lutovinov, A., Suleimanov, V., Sunyaev, R., & Zheleznyakov, V. 2004, A&A, 426, 253
Revnivtsev, M. & Sazonov, S. 2007, A&A, 471, 159
Revnivtsev, M., Sazonov, S., Churazov, E., et al. 2009, Nature, 458, 1142
Revnivtsev, M., Sazonov, S., Gilfanov, M., Churazov, E., & Sunyaev, R. 2006, A&A, 452, 169
Revnivtsev, M., Sazonov, S., Krivonos, R., Ritter, H., & Sunyaev, R. 2008, A&A, 489, 1121
Revnivtsev, M., Vikhlinin, A., & Sazonov, S. 2007, A&A, 473, 857
Ritter, H. & Kolb, U. 2003, A&A, 404, 301
Sazonov, S., Revnivtsev, M., Burenin, R., et al. 2008, A&A, 487, 509
Sazonov, S., Revnivtsev, M., Gilfanov, M., Churazov, E., & Sunyaev, R. 2006, A&A, 450, 117
Schmidt, G. D., Harris, H. C., Liebert, J., et al. 2003, ApJ, 595, 1101
Schmidt, M., Hasinger, G., Gunn, J., et al. 1998, A&A, 329, 495
Suleimanov, V., Poutanen, J., Falanga, M., & Werner, K. 2008, A&A, 491, 525
Suleimanov, V., Revnivtsev, M., & Ritter, H. 2005, A&A, 435, 191
Tanaka, T., Uchiyama, Y., Aharonian, F. A., et al. 2008, A&A, 685, 988
Tanaka, Y. 2002, A&A, 382, 1052
Tanuma, S., Yokoyama, T., Kudoh, T., et al. 1999, PASJ, 51, 161
Türler, M., Chernyakova, M., Courvoisier, T., et al. 2010, A&A, 512, A49
Ueda, Y., Takahashi, T., Inoue, H., et al. 1999, ApJ, 518, 656
Valinia, A. & Marshall, F. E. 1998, ApJ, 505, 134
Valinia, A., Tatischeff, V., Arnaud, K., Ebisawa, K., & Ramaty, R. 2000, ApJ, 543, 733
Vauclair, G., Schmidt, H., Koester, D., & Allard, N. 1997, A&A, 325, 1055
Warwick, R. S., Turner, M. J. L., Watson, M. G., & Willingale, R. 1985, Nature, 317, 218
Weisskopf, M. C., Brinkman, B., Canizares, C., et al. 2002, PASP, 114, 1
Wickramasinghe, D. T. & Ferrario, L. 2005, MNRAS, 356, 1576
Winget, D. E., Nather, R. E., Clemens, J. C., et al. 1991, ApJ, 378, 326
Worrall, D. M., Marshall, F. E., Boldt, E. A., & Swank, J. H. 1982, ApJ, 255, 111
Wu, K. 1994, Proceedings of the Astronomical Society of Australia, 11, 61
Wu, K., Chanmugam, G., & Shaviv, G. 1994, ApJ, 426, 664
Yamasaki, N. Y., Ohashi, T., Takahara, F., et al. 1997, ApJ, 481, 821
Yamauchi, S. & Koyama, K. 1993, ApJ, 404, 620

Chapter 3
Constructing an X-ray Model of Accreting White Dwarf Binaries

In this chapter, we review a geometry of mass accretion onto a magnetic white dwarf (WD), hydrostatic shock relations, and electromagnetic emission processes which take place in the accretion column. Then, an X-ray spectral model is constructed numerically to reproduce spectra of magnetic WDs. This model is utilized in Chap. 5 to estimate the gravitational potential depths of magnetic cataclysmic variables (CVs) observed in X-ray wavelength.

3.1 Improvements in This Work

Procedures for constructing an emission model of magnetic CVs, i.e., a set of physical assumptions, X-ray emission processes, equation formalism, and numerical calculation technique, were previously studied by many authors (Sect. 2.2.5; e.g., Wu et al. 1994; Cropper et al. 1998, 1999; Ramsay 2000; Suleimanov et al. 2005). In the present study, although we also generally followed those procedures, we improved the model in two aspects which were less important in the previous studies due to limited data quality. Namely, we utilized (1) the plasma cooling function with variable heavy metal abundances and (2) the X-ray spectral code which incorporates atomic X-ray emission lines from the metals. The former is considered in Sect. 3.3, and the latter in Sect. 3.5. Particularly, the atomic emission lines contained in our model, e.g., those from highly ionized Fe ions, are critical for determining a plasma temperature. As reviewed in Sect. 5.4.2, we think that the inclusion of the lines greatly improved an accuracy of our WD mass estimation compared to the previous ones.

3.2 Assumed Accretion Geometry and Boundary Conditions

As illustrated in Fig. 3.1, we assume a mass-accreting magnetic WD to have a one-dimensional cylindrical accretion column, or a post-shock region (PSR), as introduced in Sect. 2.2.3. Channeled by the magnetic field of the WD, gas is considered

T. Yuasa, Suzaku *Studies of White Dwarf Stars and the Galactic X-ray Background Emission*, Springer Theses, DOI: 10.1007/978-4-431-54219-3_3, © Springer Japan 2013

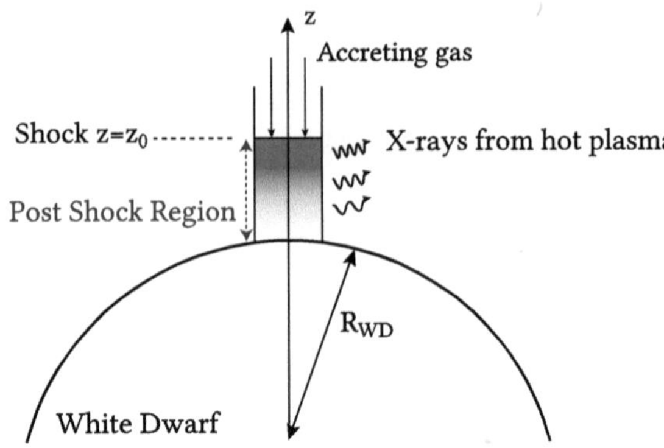

Fig. 3.1 Assumed geometry of the accretion column of a magnetic WD

to fall into the accretion column with its supersonic free-fall velocity achieved by releasing the gravitational potential energy. At a certain height, approximately several hundred kilometers from the surface (a few to several percent of the WD radius), there stands a strong shock, which converts a considerable fraction of the kinetic energy (bulk speed) of the gas into the thermal energy. Thus, after passing the shock, the gas temperature dramatically increases to 10^{7-8} K.

While the shock-heated gas falls subsonically down along the magnetic field lines, it cools by emitting X-rays. Finally, the gas is assumed to softly land onto the atmosphere of the WD. Because the magnetic fields of IPs are not as strong as those of Polars, the dominant cooling process of the heated gas is optically-thin thermal X-ray emission (free-free plus free-bound plus bound-bound), and the cyclotron cooling can be considered negligible; see for example Eq. (2) of Wu et al. (1995). Oppositely because the stronger magnetic field induces enhanced cyclotron emission, and reduces plasma temperature in Polars, their hard X-ray emission is weaker compared to that from IPs. It is also difficult to incorporate the effect of strong magnetic field into a spectral model. Therefore, we only analyze IPs in the present study. This has little effect on the GRXE study since IP dominates hard X-ray GRXE signals according to the "Point Source" scenario (Sect. 2.1.2).

Including these, the physical assumptions made in the present model are summarized below. The same assumptions were used in some previous studies (e.g., Suleimanov et al. 2005; Cropper et al. 1999).

- 1-dimensional cylindrical accretion (curvature of magnetic field neglected)
- temperature before the shock to be zero
- accreting plasma treated as ideal gas
- a strong shock
- cooling via free-free, free-bound, and bound-bound process (cyclotron cooling neglected)
- gravitational force attracts the gas in the PSR
- soft landing of cooled the gas onto the WD surface

3.3 Formulation of Conservation Equations

With the assumptions explained in the previous section, an accreting plasma flow can be described with hydrodynamical equations such as;
the mass conservation,

$$\frac{\mathrm{d}}{\mathrm{d}z}(\rho v) = 0, \tag{3.1}$$

the momentum conservation,

$$\frac{\mathrm{d}}{\mathrm{d}z}(\rho v^2 + P) = -\frac{GM_{\mathrm{WD}}}{z^2}\rho, \tag{3.2}$$

and the energy conservation,

$$v\frac{\mathrm{d}P}{\mathrm{d}z} + \gamma P\frac{\mathrm{d}v}{\mathrm{d}z} = -(\gamma - 1)\Lambda. \tag{3.3}$$

Here, z, ρ, v, P, γ, and M_{WD} are the height measured from the WD center, plasma density, velocity, pressure, the adiabatic index, and the WD mass, respectively. Λ is a quantity describing an energy loss rate via optically-thin thermal X-ray emission (see below). Since we assume an ideal gas with $\gamma = 5/3$, the equation of state of the gas is given as

$$P = \frac{\rho k T}{\mu m_{\mathrm{H}}}, \tag{3.4}$$

where k, $\mu = 0.62$, and m_{H} are the Boltzmann constant, mean molecular weight, and the hydrogen mass, respectively.

In Suleimanov et al. (2005) and Brunschweiger et al. (2009), Λ was taken from a cooling function for solar chemical composition (i.e., $Z_{\mathrm{Fe}} = 1$) calculated by Sutherland and Dopita (1993). As shown later, iron abundance is clearly smaller than unity in many IPs, and therefore it is necessary to take this into account when calculating PSR plasma profiles. Therefore, in the present study, we utilized a cooling function newly reported by Schure et al. (2009), which lists cooling rates for individual heavy elements in addition to those for hydrogen and helium. Using this, we calculated cooling rates for different (sub-solar) abundances as shown in Fig. 3.2. In higher temperatures where the cooling function of Schure et al. (2009) does not cover ($T \gtrsim 10^{8.5}$ K), we extrapolated it with a spline function.

By integrating Eq. (3.1), we introduce specific accretion rate a as

$$\rho v = a. \tag{3.5}$$

When a fraction f of the PSR cross section to the WD surface area is set, the specific accretion rate a can be related to the total accretion rate $\dot{M}$ via $\dot{M} = 4\pi R_{\mathrm{WD}}^2 f a$. Since the luminosity of a PSR is determined by liberated gravitational energy ($\propto \dot{M}$),

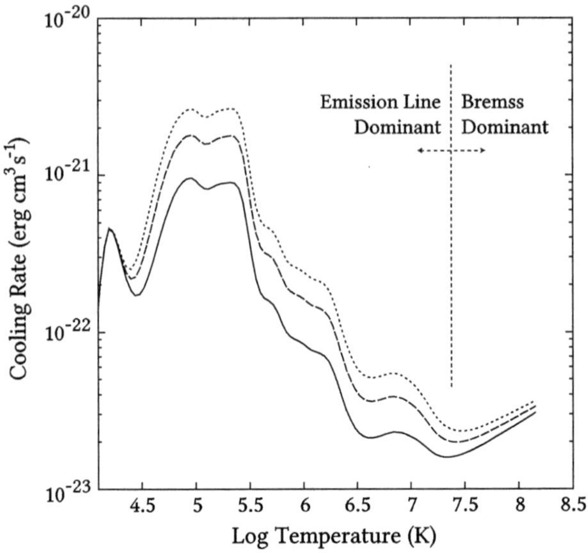

Fig. 3.2 Examples of the plasma cooling functions synthesized using the result of Schure et al. (2009) for plasmas with metal abundances of 0.5 $Z_\odot$ (*solid line*), 1.0 $Z_\odot$ (*dashed line*), and 1.5 $Z_\odot$ (*dotted line*). Atomic emission lines and electron-proton bremsstrahlung are dominant cooling processes below and above $\sim 10^{7.5}$ K

the product af can be regarded as the model normalization. Although a and f can be free parameters in general (Suleimanov et al. 2005), and the numerical solution changes as a varies, we fixed them at $a = 1.0\,\mathrm{g\,cm^{-2}\,s^{-1}}$ and $f = 0.001$ (0.1 % of the WD surface area) for the sake of easy comparison with previous reports. The assumption of f value is based on reported observational results (e.g., $f < 0.009$ by James et al. 2002 for PQ Gem; $f < 0.002$ by Hellier 1997 for XY Ari) and a theoretical constraint ($0.001 \lesssim f \lesssim 0.02$ by Rosen 1992). Instead of changing the factors a and f independently, we simply introduced a parameter which linearly adjusts the normalization of the model calculated with a and f fixed at above values. Effects of changing a on our result are discussed in Sect. 5.3.3.

Before solving the equations, we briefly check validity of the optically-thin assumption for X-ray photons emitted in a PSR. In a high-temperature plasma, electron scattering is the only possible process which acts as an opacity source for X-ray photons. The optical depth τ_T for this process can be written as

$$\tau_\mathrm{T} = n_\mathrm{e} \sigma_\mathrm{T} d, \tag{3.6}$$

where n_e, σ_T, and d are electron number density, the Thomson cross section, and typical path length in the PSR. For d, we substitute the diameter of the PSR.

$$d = (4f)^{1/2} R_\mathrm{WD}. \tag{3.7}$$

The electron density in the PSR is calculated using the bulk (falling) velocity of the plasma v as

$$n_{\mathrm{e}} \sim \frac{a}{v \mu m_{\mathrm{H}}}. \tag{3.8}$$

By considering that the shock height is relatively small compared to the WD radius, we approximate v at the free-fall velocity v_{ff}.

$$v = v_{\mathrm{ff}} = \sqrt{\frac{2GM_{\mathrm{WD}}}{R_{\mathrm{WD}}}} = 3.7 \times 10^{8} \left(\frac{M_{\mathrm{WD}}}{0.5\,M_{\odot}}\right)^{1/2} \left(\frac{R_{\mathrm{WD}}}{10^{9}\ \mathrm{cm}}\right)^{-1/2} \mathrm{cm\ s^{-1}}. \tag{3.9}$$

Substitution of assumed parameters, such as $a = 1.0\ \mathrm{g\ cm^{-2}\ s^{-1}}$ and $f = 0.001$, we obtain

$$\tau_{\mathrm{T}} = 0.11 \left(\frac{a}{1.0\ \mathrm{g\ cm^{-2}\ s^{-1}}}\right) \left(\frac{f}{0.001}\right)^{-1/2} \left(\frac{M}{0.5\,M_{\odot}}\right)^{-1/2} \left(\frac{R}{10^{9}\ \mathrm{cm}}\right)^{-1/2} \tag{3.10}$$

Thus, the optical thickness is smaller than unity in typical parameter values, and the plasma can be considered as optically thin for X-ray photons.

3.4 Numerical Implementation

After combining Eqs. (3.2) and (3.3) using Eq. (3.5), to find the solution of these equations reduces to solving an initial value problem of ordinary differential equations, and therefore, rather straightforward. As explained, for example, in Cropper et al. (1999) and Suleimanov et al. (2005), this can be practiced with the so-called shooting algorithm. Each trial starts from an assumed shock height $z = z_0$, where the initial parameters are set as follows according to the strong shock limit of the Rankine-Hugoniot relation (the same boundary conditions as used in Suleimanov et al. 2005);

$$v_0 = \frac{1}{4}\sqrt{\frac{2GM_{\mathrm{WD}}}{z_0}}, \tag{3.11}$$

$$\rho_0 = \frac{a}{v_0}, \tag{3.12}$$

$$P_0 = 3 a v_0, \tag{3.13}$$

$$T_0 = \frac{3\mu m_{\mathrm{H}}}{k} v_0^2. \tag{3.14}$$

Equation (3.11) is derived from the strong-shock assumption. The lower limit of the calculation, $z = R_{\mathrm{WD}}$, is specified by the WD mass-radius relation of Eq. (2.10).

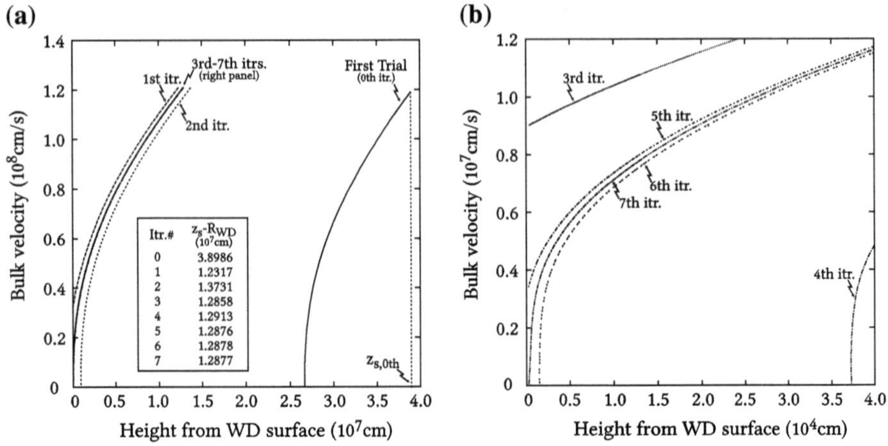

Fig. 3.3 **a** An example of a vertical velocity structure, or bulk velocity of plasma $v(z)$, in the PSR iteratively calculated using the shooting algorithm. The equations with typical input parameters ($M_{\rm WD} = 0.7\ M_{\odot}$ and $Z_{\rm Fe} = 0.5\ Z_{\odot}$) are utilized for this calculation. It is hard to recognize differences of solutions for the third to seventh iterations in the plot (see *panel* **b**). **b** A close-up view of the same plot for showing behavior of the third to seventh iterations near the WD surface. Note that all velocity profiles except for that of the seventh iteration (the final solution) do not fulfill the imposed soft-landing condition, i.e., $v = 0$ at the WD surface, and therefore, they are not the final solution

At the WD surface, another boundary condition of $v = 0$ is imposed to fulfill the soft-landing assumption.

The equations are solved from the initial point ($z = z_0$; shock height), and deviations of each quantities at the final point ($z = z_{\rm WD} = R_{\rm WD}$; WD surface) from the assumed boundary conditions are examined. By taking into account the deviation as a feedback to a new initial z_0, the equations are solved again. Until the feedback becomes small enough, below 10^{-4} of derived z_0 in the present case, this procedure is repeated. Practically, this method converged in several to a few tens of iterations at each value of $M_{\rm WD}$.

Figure 3.3 presents a schematic example of this procedure showing vertical velocity structure, i.e., $v(z)$, in the PSR. The first trial was computed with an initial shock height which was set arbitrarily at $z_{\rm s} = 1.05 R_{\rm WD}$ (a shock height of 5 % of the WD radius). A numerical evaluation of each equation starts from right to left (top of the PSR to bottom), and stops at $z = R_{\rm WD}$, i.e., the surface of the WD. In this example, after seven times of feedbacks, the initial value and resulting velocity profile met the convergence condition providing a solution $z_{\rm s} = 1.2877 \times 10^7\ {\rm cm} + R_{\rm WD} = 1.0165 R_{\rm WD}$.

We executed the calculation over a range of $0.20\ M_{\odot} \leq M_{\rm WD} \leq 1.39\ M_{\odot}$ with a step of $0.01\ M_{\odot}$. Figure 3.4 shows the dependency of the shock height and the shock temperature on the WD mass by interpolating computed data points with spline functions. A fact that the resulting shock temperature exceeds 10 keV, or equivalently

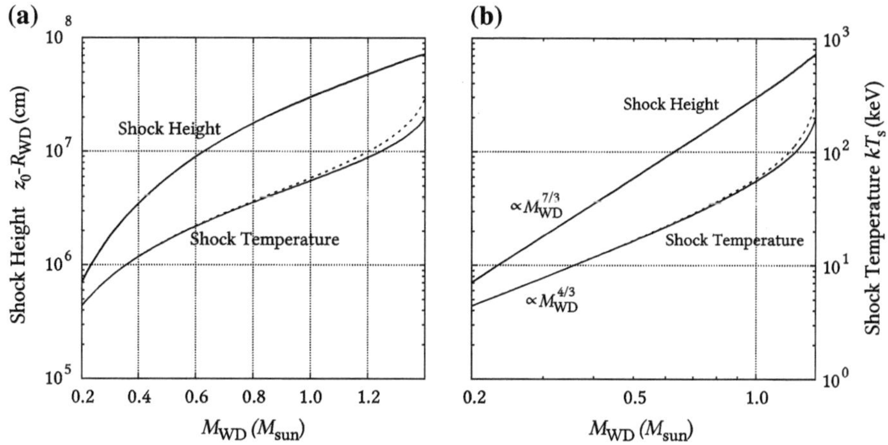

Fig. 3.4 **a** Results of the numerical solutions for the shock height from the WD surface (*thick solid line*) and the shock temperature (*thin solid line*), shown against the WD mass. For comparison, the *dashed line* shows the shock temperature calculated by assuming no-gravity in the PSR. **b** A log-log plot of the same curves shown in *panel* (**a**)

1.2×10^8 K, in IPs should be emphasized because in such a high temperature a hot plasma emits characteristically hard X-rays making themselves conspicuous in the hard X-ray sky. In Fig. 3.4, as heavier mass as a WD gets, the gas velocity at the shock height becomes higher, and therefore the gas density falls for a given accretion rate. This causes a decrease in the plasma cooling rate and cooling time, yielding taller shock heights.

In Fig. 3.5, we present profiles of the temperature and density calculated for an IP with a WD mass of 0.7 $M_\odot$, for comparison with the results of Suleimanov et al. (2005). The shock height of 0.0165 R_{WD}, which we derived, is somewhat different from that of Suleimanov et al. (2005), 0.018 R_{WD}. This differences might arise from difference of the cooling functions and implementations of the shooting algorithm. In the present study, we do not investigate this difference since the shock temperatures in the two cases differ only by $\sim$0.3 %.

Cropper et al. (1999) showed that an inclusion of the gravitational attraction changes WD mass estimations especially in high mass ($\sim$1.0 $M_\odot$) systems. This can be confirmed with the comparison of the profiles of shock temperature calculated with and without gravitational attraction as shown in Fig. 3.4. For example, at $M_{WD} = 1.2\ M_\odot$, the no-gravity profile (dashed curve) overestimates a shock temperature and following M_{WD} by $\sim$10 % compared with the present result. Note that in Cropper et al. (1999), they used an analytic free-free cooling rate (Wu 1994) which is described in their Appendix A. Therefore, the resulting shock temperatures (their Fig. 1) are slightly different from our values (Fig. 3.4).

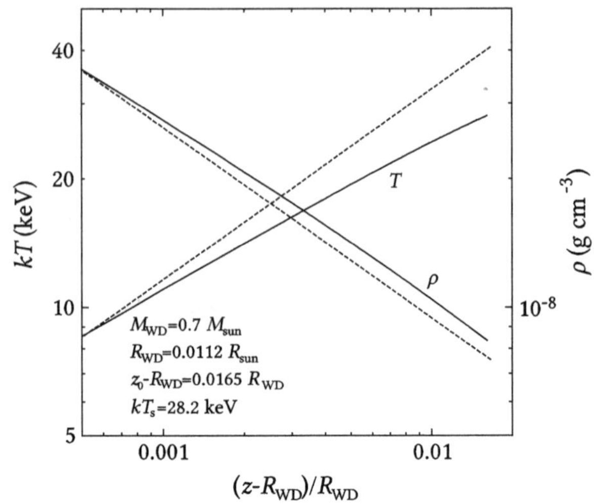

Fig. 3.5 Temperature and density profiles of a PSR calculated for an IP of $M_{\rm WD} = 0.7\ M_{\odot}$ and $Z_{\rm Fe} = 0.5\ Z_{\odot}$. For easy comparison with Suleimanov et al. (2005), the plot employs the same parameter values and the same style as their Fig. 2. Labels give the WD radius calculated using the mass-radius relation of Eq. (2.10), as well as the shock height ($z_0 - R_{\rm WD}$) and the shock temperature ($kT_{\rm s}$) from Fig. 3.4. For comparison, power-law functions with indices of $+1/2$ and $-1/2$ are overlaid as *dashed lines* for the temperature and the density profiles, respectively

3.5 Composing an Emerging X-ray Spectrum

Based on the calculation results, we constructed, for each $M_{\rm WD}$, an emerging spectrum of the PSR. Since plasmas in individual strata of a PSR emit X-rays with different spectral shapes depending on their temperatures, we have to convolve temperature (and density) profiles of the PSR with a single temperature plasma emission model. Although the model of Suleimanov et al. (2005) calculated only bremsstrahlung continuum using an analytical formula and did not take into account atomic emission lines, we have to calculate them properly. This is because *Suzaku* can resolve, particularly, the three Fe lines in the 6–7 keV band, and they are very important tools for plasma diagnostics.

Therefore, we replaced the simple bremsstrahlung continuum with an optically-thin emission code `APEC` (Smith et al. 2001) which includes emission lines, and constructed a model for the summed X-ray spectrum. In this way, we prepared a series of convolved model spectra for each $M_{\rm WD}$ by changing the metal abundance of the `APEC` model from 0.1 to 2.0 $Z_{\odot}$ with a step of 0.1 $Z_{\odot}$.

Because the currently available version of `APEC` does not calculate the spectrum above $E > 50$ keV, we can miss the exponential cutoff feature, which is crucial in the calculation of high temperature layers ($kT > 20$–30 keV). Although this energy range is not directly employed in our data analysis, the model shapes in that range are considered important, since the HXD response is not diagonal, and photons with

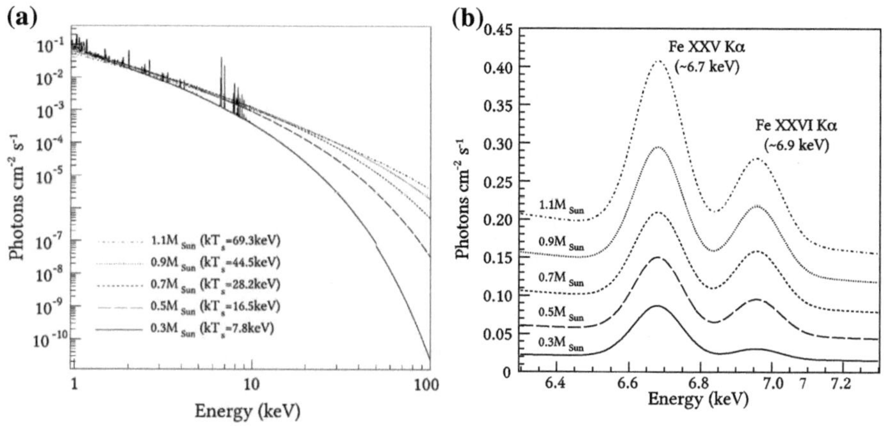

Fig. 3.6 **a** Composed emerging spectra for IPs with $M_{\rm WD} = 0.3, 0.5, 0.7, 0.9$, and $1.1\ M_\odot$, with an Fe abundance of $0.5\ Z_\odot$ and a mass accretion rate of $a = 1\ {\rm g\ cm^{-2}\ s^{-1}}$, observed at a distance of 100 pc (a typical distance to sources analyzed in the present study). The ordinate is a predicted photon flux at the observing point. Discontinuity seen in the spectrum of $0.3\ M_\odot$ at 50 keV is due to the two different models used in the spectral synthesis for above and below the energy (see text). **b** The same plot but the Fe line complex in the 6–7 keV band is enlarged in linear scales. For better visibility, spectra are smoothed with a typical energy resolution of the XIS (150 eV FWHM), and then shifted vertically by multiplying a factor of 1.5 for every step from 0.3 to $1.1\ M_\odot$

$E > 50$ keV sometimes produce signals in significantly lower energies. Therefore, above $E > 50$ keV, we substituted spectra estimated by `meka` model (which provides a spectrum in that energy range), with its normalization set 5.5 % smaller than that of the corresponding `APEC` model. This renormalization is to make the two spectra match smoothly at $E = 50$ keV under high temperatures where this substitution is of particular importance.

Figure 3.6 shows synthesized emerging spectra for various WD masses with $Z_{\rm Fe} = 0.5\ Z_\odot$. In lighter WD masses (e.g., $\lesssim 0.7\ M_\odot$), a feature of an exponential cutoff can be seen clearly in the spectra, especially in the HXD/PIN energy band (10–50 keV). The Fe emission line complex in the 6–7 keV is clearly resolved by the XIS (Fig. 3.6 panel b). In particular, the 6.9 keV line which originates from highly ionized Fe XXVI (i.e., H-like Fe) grows as the shock temperature increases (higher $M_{\rm WD}$). These spectral properties again ensure that the simultaneous observation of IPs by the XIS and the HXD are suitable for performing detailed spectroscopies and WD mass estimations.

When $Z_{\rm Fe}$ is changed, intensities of emission lines increase linearly relative to the continuum which is almost intact. Figure 3.7 exemplifies this, showing spectra of an IP with $M_{\rm WD} = 0.7\ M_\odot$ calculated for various abundance.

Figure 3.8 compares spectra of the composed IP model and a single-temperature optically-thin thermal model (`APEC`) of which plasma temperature is fixed at the shock temperature of the IP model. From this, it can be recognized that the IP model has higher fluxes in soft X-ray band, and therefore, lower "color temperature" than that of the single-temperature model.

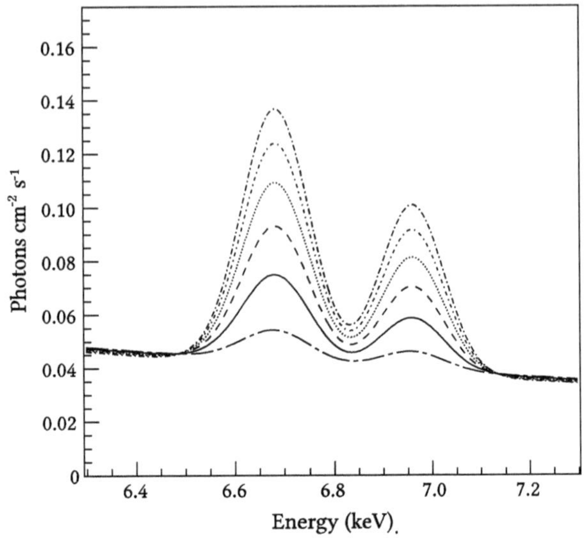

Fig. 3.7 The Fe line complex of the composed emerging spectra for IPs with $M_{\rm WD} = 0.7\ M_\odot$ calculated under the same condition used in Fig. 3.6 but with $Z_{\rm Fe} = 0.1$, 0.3, 0.5, 0.7, 0.9, and 1.1 $Z_\odot$ (from *bottom* to *top*). Note that the *dashed curve*, i.e., the third lowest one, in this plot is equivalent to the dashed one presented in Fig. 3.6

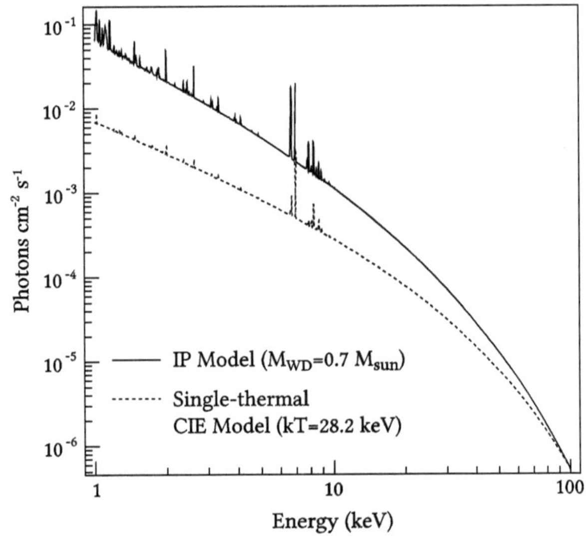

Fig. 3.8 A comparison of the composed IP spectrum (*solid curve*) and the single-temperature CIE model (*dashed curve*). A WD mass of 0.7 $M_\odot$ is assumed in the composed model, and therefore this is the same as one shown in Fig. 3.6 panel a. A plasma temperature of the CIE model was set at the shock temperature $kT_{\rm s}$ of this WD mass. The normalization of the CIE model was adjusted so that its flux matches that of the IP model at 100 keV

Finally, a set of total spectra calculated in this way were converted to a so-called *local-model* file of the spectral analysis software XSPEC (Arnaud 1996). The free parameters of the model are the WD mass, the metal abundance, and the model normalization. In this local model, spectra for masses and abundances which are in between the computation grid points are generated by interpolating spectra of the surrounding 4 grid points. Note that since we performed the model calculation with sufficiently fine grid intervals, the interpolation does not have any systematic effect on the result.

References

Arnaud, K. A. 1996, in Astronomical Society of the Pacific Conference Series, Vol. 101, Astronomical Data Analysis Software and Systems V, ed. G. H. Jacoby & J. Barnes, 17
Brunschweiger, J., Greiner, J., Ajello, M., & Osborne, J. 2009, A&A, 496, 121
Cropper, M., Ramsay, G., & Wu, K. 1998, MNRAS, 293, 222
Cropper, M., Wu, K., Ramsay, G., & Kocabiyik, A. 1999, MNRAS, 306,684
Hellier, C. 1997, MNRAS, 291, 71
James, C. H., Ramsay, G., Mark Cropper, M. C., & Branduardi-Raymont, G. 2002, MNRAS, 336, 550
Ramsay, G. 2000, MNRAS, 314, 403
Rosen, S. R. 1992, MNRAS, 254, 493
Schure, K. M., Kosenko, D., Kaastra, J. S., Keppens, R., & Vink, J. 2009, A&A, 508, 751
Smith, R. K., Brickhouse, N. S., Liedahl, D. A., & Raymond, J. C. 2001, ApJ, 556, L91
Suleimanov, V., Revnivtsev, M., & Ritter, H. 2005, A&A, 435, 191
Sutherland, R. S. & Dopita, M. A. 1993, ApJS, 88, 253
Wu, K. 1994, Proceedings of the Astronomical Society of Australia, 11, 61
Wu, K., Chanmugam, G., & Shaviv, G. 1994, ApJ, 426, 664
Wu, K., Chanmugam, G., & Shaviv, G. 1995, ApJ, 455, 260

References

Chapter 4
Instrumentation

In the present thesis, the spectral model constructed in Sect. 3.5 is applied to X-ray spectra of magnetic cataclysmic variables (CVs) for estimating their white dwarf (WD) masses. This chapter describes instruments used to obtain the spectra and the computer software employed to analyze the acquired data.

4.1 The *Suzaku* X-ray Observatory

The *Suzaku* satellite (Mitsuda et al. 2007) is the fifth Japanese space X-ray observatory which was launched on 2005 July 10 from Uchinoura, Kagoshima, Japan. It has a near-circular low-Earth orbit with an altitude of 570 km and an orbital period of about 96 min. There are two main X-ray sensors onboard the satellite; the X-ray Imaging Spectrometer (XIS; Koyama et al. 2007), and the Hard X-ray Detector (HXD; Takahashi et al. 2007; Kokubun et al. 2007). Although *Suzaku* carried another detector, the X-Ray Spectrometer (XRS) which aimed at very high resolution X-ray spectroscopy, it became inoperable due to unexpected evaporation of liquid helium coolant which occurred in the early commissioning phase.

Figure 4.1 shows in-orbit configuration of the satellite and internal layout of individual subsystems. In Table 4.1, basic parameters of *Suzaku* and its onboard detectors are listed. Compared to other X-ray missions, *Suzaku* has two outstanding features; a wide-band energy coverage (0.2–600 keV), and low instrumental backgrounds. The present thesis makes a full use of these features of *Suzaku*.

To maximize these advantages, *Suzaku*, with its rather narrow field of view, performs a pointing observations on selected celestial targets instead of surveying the sky. By such an observation, we aim to obtain wide-band X-ray spectra of the targets within moderate exposures, e.g., several hours to a few days, mainly depending on the source intensities. Because of the low-Earth orbit, an observation is periodically interrupted for several tens of minutes typically every orbit due to Earth occultations of the target, or passages through a region of very high particle background above the South Atlantic Ocean, called South Atlantic Anomaly (SAA). Effectively, the net exposure becomes about 50 % of the real time due to these interruptions.

T. Yuasa, Suzaku *Studies of White Dwarf Stars and the Galactic X-ray Background Emission*, Springer Theses, DOI: 10.1007/978-4-431-54219-3_4, © Springer Japan 2013

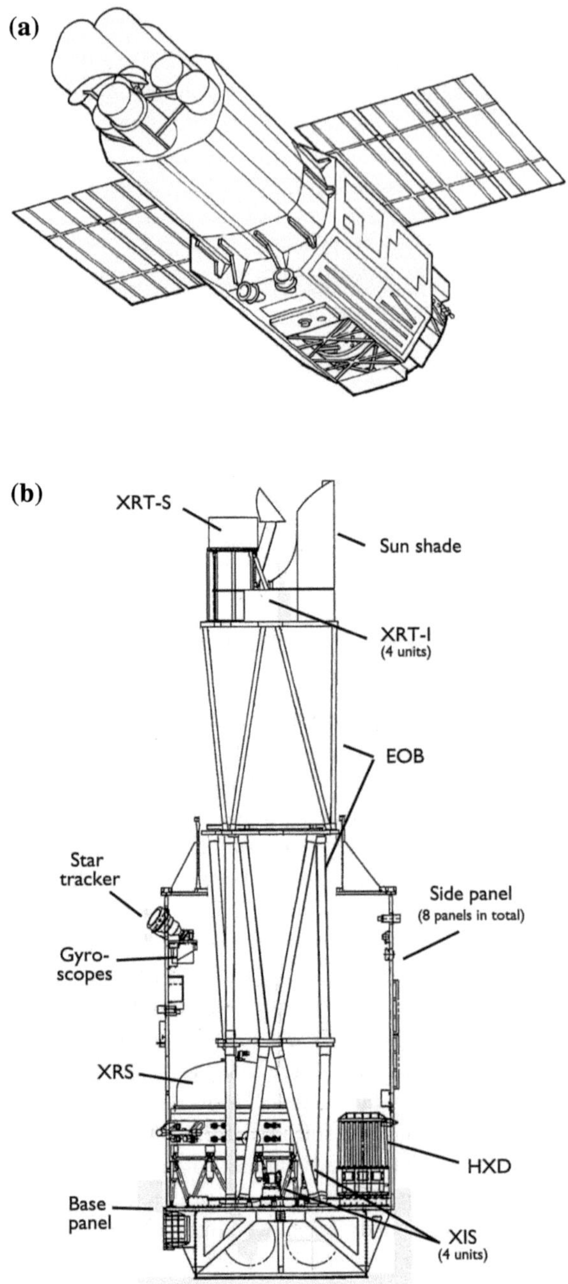

Fig. 4.1 **a** The in-orbit configuration of *Suzaku*. **b** A side view of the satellite, with the side panels and the solar paddles both removed. Adopted from Mitsuda et al. (2007) with permission of the author and the publisher

Table 4.1 Basic parameters of *Suzaku*, and its on-board instruments

Space craft	Altitude	~570 km
	Orbital inclination	31°
	Orbital period	96 min
	Weight	1706 kg (at launch)
	Length	6.5 m (in-orbit configuration)
	Science instruments	4 × (XRT + XIS), HXD
XRT	Focal length	4.75 m
	Effective area	440 cm^2 at 1.5 keV, 250 cm^2 at 8 keV
	Angular resolution	2′ (half power diameter)
XIS	Energy coverage	0.2–12 keV
	Field of view	18′ × 18′
	Time resolution	8 s (normal mode), 7.8 ms (P-sum mode)
HXD	Energy coverage	12–70 keV (PIN), 50–600 keV (GSO)
	Field of view	4°.5 × 4°.5
	Effective area	~160 cm^2 at 20 keV, ~260 cm^2 at 100 keV
	Time resolution	61 μs

4.1.1 The X-ray Imaging Spectrometer and the X-ray Telescope

The X-ray Imaging Spectrometer (XIS; Koyama et al. 2007) is a low-energy imaging spectrometer. It consists of four X-ray imaging charge coupled device (CCD) cameras, working as focal plane detectors of the X-ray Telescope (XRT; Serlemitsos et al. 2007).

The XRT consists of five sets of a Wolter-I type X-ray reflection mirror each with a focal length of 4.75 m. Four of them, called XRT-I modules, are coupled to the four XIS cameras, whereas the other (XRT-S) is for the XRS. One of the four XRT-I mirrors is shown in Fig. 4.2. Each mirror is composed of a large number of nested conical reflectors, each reflecting X-ray photons when their grazing incident angle is $\lesssim 0.7°$. One XRT-I module consists of 1400 reflectors made of Aluminium foils, of which the surface is coated with Au. The XRT has strong points in the light weight (19.3 kg/telescope) and the large effective area (see below) against telescopes used in *Chandra* (950 kg) and *XMM-Newton* (500 kg). However, the angular resolution of the XRT is worse compared to these by about a factor of 100 and 20.

Figures 4.3 and 4.4 show a schematic drawing of the XIS sensor and the effective area of the XRT+XIS, respectively. The CCD chip detects X-ray photons collected by the XRT, and records their individual positions and energies in a photon-counting mode. Three (XIS0, 2, and 3) use front-side illuminated (FI) CCD chips, while the other does back-side illuminated one (BI; XIS1). The difference of the illumination direction changes the quantum efficiency and effective area of the chip, especially at lower energies (e.g., <2 keV).

The energy coverage of the XRT+XIS is nominally 0.2–12 keV, and the energy resolution is roughly 160–180 eV at the 5.9 keV calibration point. Since 2006 October,

Fig. 4.2 A picture of an XRT-I. Adopted from Serlemitsos et al. (2007) with permission of the author and the publisher

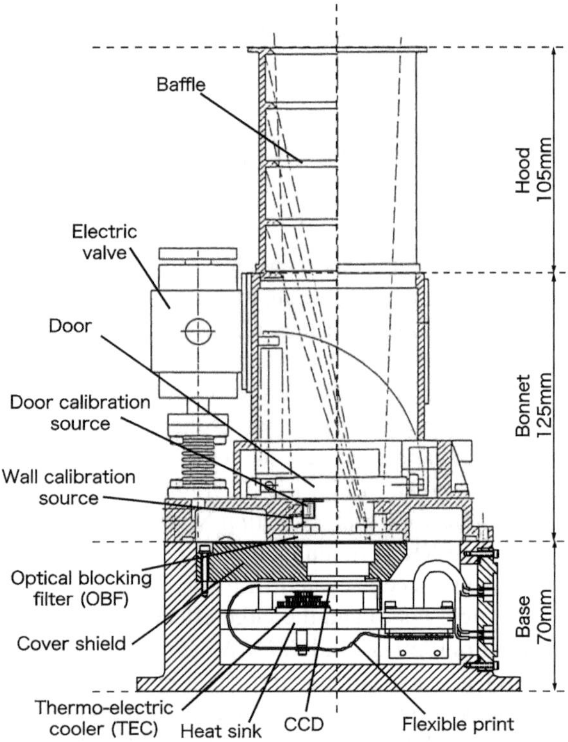

Fig. 4.3 A cross section of the XIS sensor. The CCD chip is located in the base structure, and the baffle is placed on it to reduce background photons from outside the field of view. Incoming X-rays collected by the XRT come down from the top. Adopted from Koyama et al. (2007) with permission of the author and the publisher

the charge-injection technique (Nakajima et al. 2008; Uchiyama et al. 2009) was applied to the CCDs to restore the energy resolution, which kept degrading after the launch due to radiation damage. Time dependent change of the energy resolution are taken into account by the analysis software. With the energy resolution of the

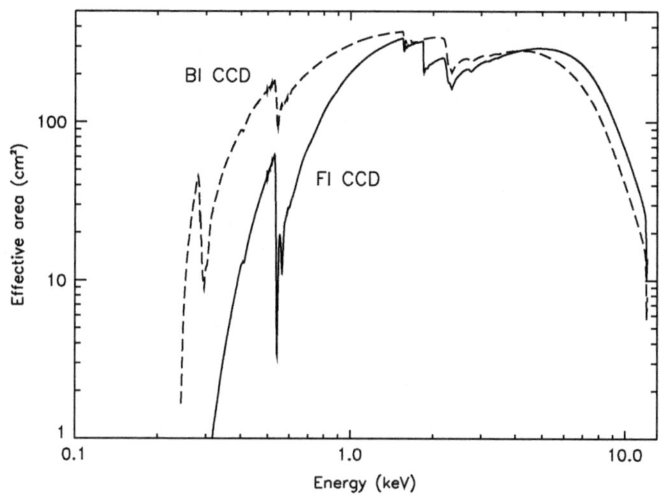

Fig. 4.4 The effective area of the XRT mirrors, when the quantum efficiency of the XIS is taken into account. Adopted from Mitsuda et al. (2007) with permission of the author and the publisher

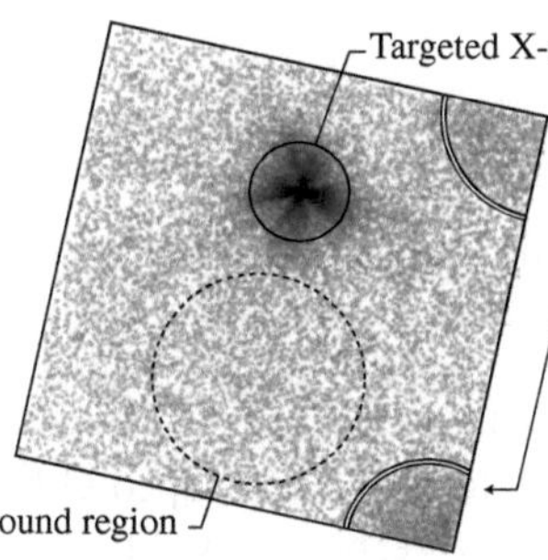

Fig. 4.5 An example image of the XIS containing a point-like X-ray source. The whole image covers a sky region of $18' \times 18'$

XIS, atomic emission lines from ionized heavy metals, which are one of characteristic features of X-ray emitting plasmas, can be resolved moderately, allowing us to distinguish ionization states of the metals. In the present study, this information is crucial to estimate a plasma temperature.

Figure 4.5 presents an example of an X-ray image taken with the XIS. Reading out the whole CCD area (1024×1024 pixels) takes 8 s, and therefore a single exposure (i.e., timing resolution) of the XIS is 8 s in the normal observation mode. Since read-out timing is artificially determined and controlled by read-out electronics, there is no "trigger"-induced operation in the XIS (c.f. self-triggered event acquisition of the HXD; see the next section). Although there are special observation modes in which faster read-outs can be performed by discarding some part of the imaging area, they were not utilized in the observations of the targets analyzed in the present thesis.

When analyzing the XIS data of a virtually point-like celestial target (i.e., not spatially extended beyond the angular resolution of the XRT), signals can be extracted from a selected region of the CCD chip. By doing so, the instrumental background can be reduced and the signal-to-noise ratio is improved dramatically. The instrumental background can be estimated from CCD regions other than that used for the source spectral extraction (Fig. 4.5). Typically, the background is at a level of $\sim 1\,\%$ of count rates of the celestial targets analyzed in the present thesis. Thus, a systematic uncertainty of the instrumental background subtraction becomes much smaller, and essentially, negligible.

When observing largely extended X-ray sources, like the Galactic Ridge X-ray Emission (Sect. 2.1), no region can be used for the background estimation because the whole CCD area contains celestial signals. In these cases, the instrumental background is estimated using a data base of the background integrated over those periods when the target is occulted by the night Earth (Tawa et al. 2008).

Since 2006 November, one of the XIS cameras, XIS2, has stopped working supposedly due to a micro-meteoroid impact. Therefore, for observations performed after that date, we utilize data taken with the other three XIS cameras.

4.1.2 The Hard X-ray Detector

The Hard X-ray Detector (HXD) is a non-imaging collimated X-ray spectrometer that consists of a 4×4 array, i.e., 16 in total, of the counter units, achieving the effective area in the hard X-ray band as plotted in Fig. 4.6. As illustrated in Fig. 4.7, each counter unit has two main detection parts; silicon *p*-intrinsic-*n* diodes (hereafter PIN) and gadolinium silicate scintillation crystals (GSO; Ce doped Gd_2SiO_5), covering the energy ranges of 10–70 and 50–600 keV, respectively. Photo absorption is the desirable interaction through which these sensors detect photons.

The PIN diodes and GSO crystals are located at the bottom of a well-type active shield. As shown in Fig. 4.7, the active shield are made of bismuth germanate (BGO; $Bi_4Ge_3O_{12}$) scintillation crystal, and reduces the background signals caused by cosmic-ray particles and cosmic X-ray background by actively tagging such signals. To further improve the signal-to-noise ratio by reducing the cosmic X-ray background contribution in the PIN energy range, a passive collimator made of phosphor bronze, which narrowers the field of view to $34' \times 34'$ (FWHM), is inserted into the well-type active shield and placed in front of PIN. Above 70 keV, the stopping power of the fine collimator gradually decreases, and the field of view becomes loosely collimated. At higher energies above $\sim$150 keV, not the fine collimator but the BGO active shield constrains the field of view to $8^\circ \times 8^\circ$. Thus, a counter unit is composed of 4 PIN diodes, 4 GSO crystals, the BGO active shield surrounding them, and 4 fine collimators, as well as a photomultiplier tube (PMT) to read scintillation pulses from GSO and BGO, and 4 charge-sensitive amplifiers for the PIN diodes.

Since the GSO crystal is viewed through the BGO shield by the attached PMT, it is crucial to distinguish scintillation signals from the two crystals. To enable this, one

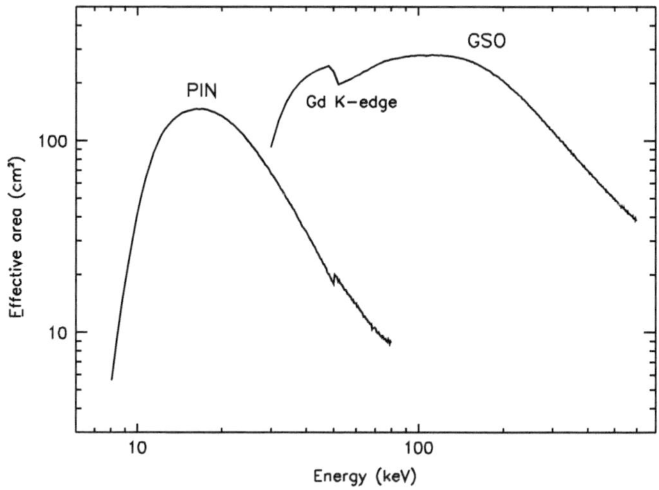

Fig. 4.6 The on-axis effective areas of the PIN and GSO sensors of the HXD, summed over the 16 well counter units. Adopted from Takahashi et al. (2007) with permission of the author and the publisher

PMT output is fed to two amplifiers with different band-pass functions each adapted for scintillation decay timescales of GSO (120 ns at −20 °C) and BGO (700 ns). When an X-ray photon is absorbed in the GSO crystal, the amplifier for the GSO (i.e., one passes higher frequencies) produces larger output voltage than the other, and vice versa.

In contrast to the time-divided read out of the XIS (Sect. 4.1.1), the HXD processes X-ray (and particle background) signals on a event-by-event basis to enable anti-coincidence using the BGO shield, and to endure high background count rate caused by cosmic rays. When one of the PIN diodes, the GSO crystals, or the BGO active shields detects an X-ray photon or a cosmic-ray particle, a trigger signal is generated by a comparator in the electronics, and signal outputs from all sensors including the PIN diodes, and the PMT (i.e., GSO and BGO), are recorded. Onboard hardware and software processes individual event hit patterns, and accept them only when an event seems to be caused by an X-ray photon. Events with background-like hit patters are discarded onboard and in the off-line processing on ground; e.g., we can securely discard simultaneous multiple hits to the BGO active shield of neighboring counter units as shown in Fig. 4.8 case 1, because only high energy cosmic-ray particles (mostly protons) can produce such a hit pattern. Since the BGO active shield covers large solid angle around the main sensors (PIN and GSO), it also detects X-ray photons scattered at them (Fig. 4.8 case 2). Even when the incident Compton scattering event triggers an event acquisition, it can be rejected by knowing that the scattered photon also triggered the BGO active shield (i.e., PMT output). Thus, the BGO active shield plays an important role in reducing the instrumental background.

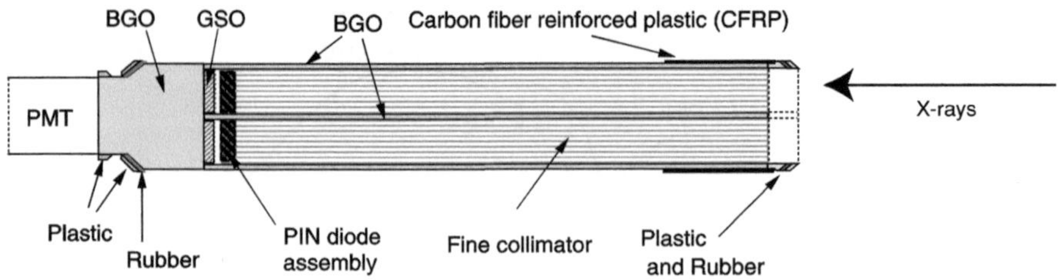

Fig. 4.7 A cross-sectional drawing of one of the 16 well-type counter units of the HXD. The *arrow* on the *right* indicates the incoming X-ray direction. Adopted, with some modifications, from Takahashi et al. (2007) with permission of the author and the publisher

Fig. 4.8 Two example cases of background events which can be rejected thanks to the BGO active shield; (1) a proton penetrating multiple counter unit, and (2) a celestial X-ray photon being scattered at the PIN diode and then absorbed by the shield

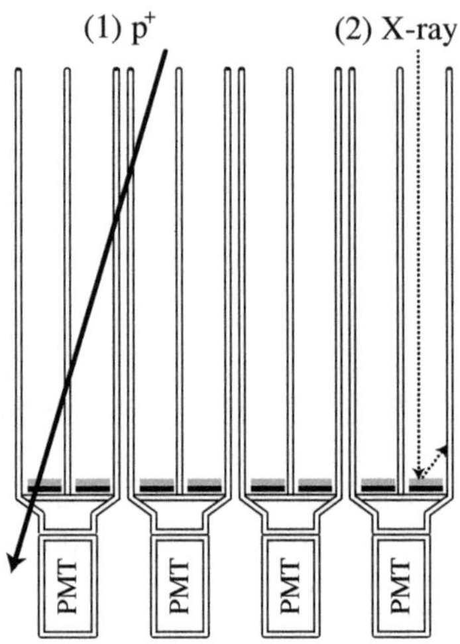

The high sensitivity in the hard X-ray energy band is crucial for detailed spectroscopy of high temperature plasmas, because their spectra have characteristic thermal or exponential cutoff features just in the energy range covered by PIN. Owing to the active anti-coincidence technique, the onboard/offline background rejection mechanism, and the tightly collimated field of view, the PIN diodes achieved the low background level, $\sim$0.5 counts s^{-1} in the 10–70 keV band (Kokubun et al. 2007), which is the lowest compared to other satellite-borne hard X-ray detectors when normalized by the effective areas. This residual background mostly consists of signals caused by (1) elastic collisions with the PI diodes and albedo neutrons created via interactions with primary cosmic-ray particles with the Earth's atmosphere (Fukazawa et al. 2009), and (2) decays of natural and cosmic-ray-induced radio isotopes. In HXD/PIN, the former dominates the residual background counts. Figure 4.9 shows typical temporal variations of the HXD/PIN background. Since the flux of the albedo neutrons depends on that of the primary cosmic rays which changes according to the local magnetic field strength (i.e., the cut-off rigidity for cosmic rays), and

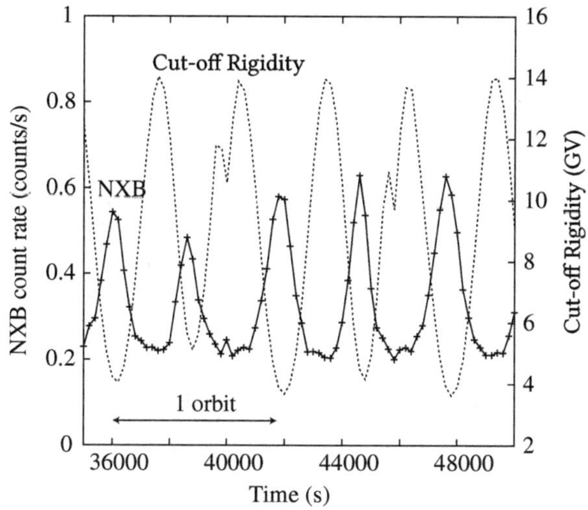

Fig. 4.9 An example of time variations of the HXD/PIN non-X-ray background (*crosses* with interpolating *solid curve*). *Dashed curve* represents the cut-off rigidity measured at the satellite position

hence, an anti-correlation between the background count rate and the cut-off rigidity is clearly seen in actual data as plotted in Fig. 4.9.

The strategy of the HXD is to observe a target continually with the collimated field of view without interleaving background (blank-sky) observations. This is to achieve a longer exposure onto the target, based on an assumption that the detector background can be predicted without its actual measurement. When analyzing observed data, the instrumental background signals must, therefore, be modeled and subtracted from the observed ones. The HXD team constructed a model for estimating the background by fitting the background count rates when targets are occulted by the Earth, and has been releasing the background event lists simulated using the model (Fukazawa et al. 2009).

The high accuracy of the background estimation is crucial in achieving the high sensitivity, and its 1σ systematic error is currently known to be $\sim$1 % (Fukazawa et al. 2009) in the PIN energy band as shown in Fig. 4.10. Practically, the 1σ systematic error of the background estimation accounts approximately for 10 % of the count rates of targets analyzed in the present thesis.

In Fig. 4.12, we visually compare the size of the fields of view of the HXD and the XIS together with the angular size of the Earth's Moon. Since the HXD is a collimator instrument, its effective area becomes the maximum when an X-ray source is located at the center of the field of view. At off-axis angles, the effective area decreases as shown in Fig. 4.11. The full width at half maximum of the angular response function is $34'$; when an X-ray source is located $34'$ away from the HXD optical axis, the effective area for the source is halved.

When observing an extended emission, for example supernova remnants or the Galactic Ridge X-ray emission (GRXE), which spreads over the HXD field of view, the surface brightness becomes convolved with the angular transmission, to constitute

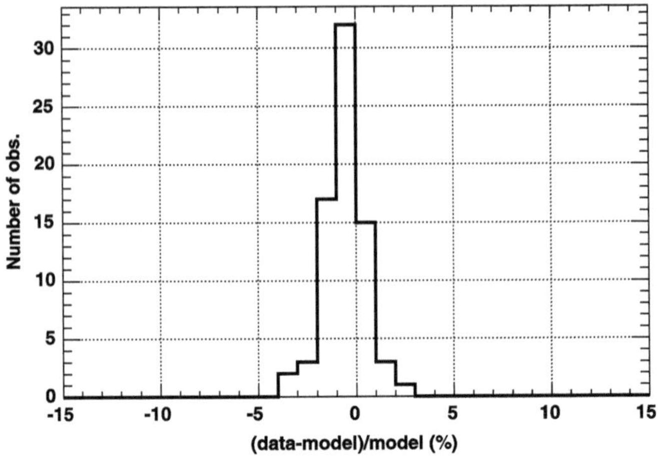

Fig. 4.10 Fractional residuals of the 15–40 keV PIN count rates obtained by subtracting the modeled instrumental background from the data actually observed during the Earth occultation of a celestial target. The integration time (i.e., exposure) of 40 ks is assumed in the model calculations. Adopted from Fukazawa et al. (2009) with permission of the author and the publisher

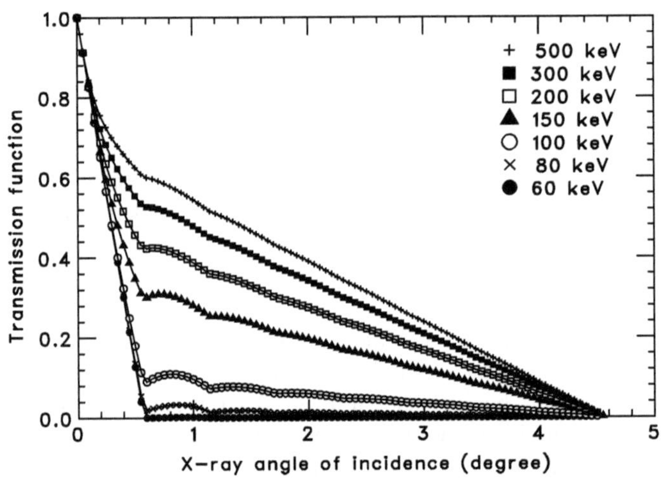

Fig. 4.11 Angular transmission functions of the HXD, defined by the fine collimators ($0.57° = 34'$ FWHM) and the BGO active shield collimation (4° FWHM). The abscissa is an angular distance from the center of the field of view. Data points of 60 keV (*filled circles*) represent the HXD/PIN angular response, and remaining for GSO. Adopted from Takahashi et al. (2007) with permission of the author and the publisher

the observed data. For such a diffuse celestial source, the effective sky coverage of the HXD becomes $34' \times 34' = 1156\,\mathrm{arcmin}^{-2}$.

Although the GSO sensor of the HXD expands the spectral coverage toward higher energies, it can detect only bright sources, due to a higher instrumental background

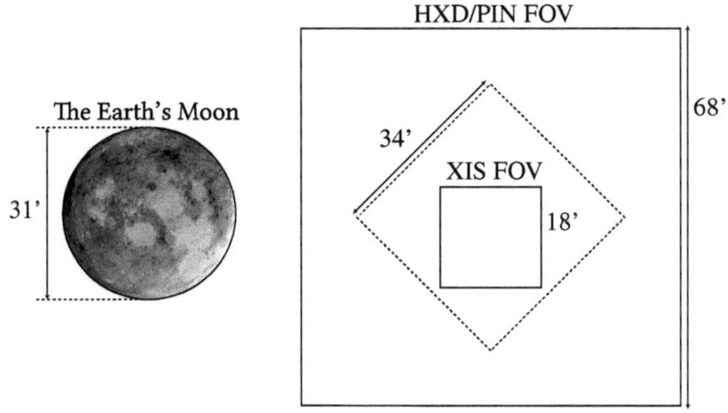

Fig. 4.12 The sky coverages of the HXD and the XIS (*solid rectangles*) compared with the angular size of the Earth's Moon ($\sim 31'$ in diameter). *Dashed rectangle* represents a contour where the transmission of the HXD fine collimator becomes 50 %. The image of the Moon was taken from the NASA Apollo web page (http://nssdc.gsfc.nasa.gov/planetary/lunar/)

(Kokubun et al. 2007; Fukazawa et al. 2009) caused by the in-orbit activation by protons. The signal count rates of celestial targets of the present thesis are far below the sensitivity limit of GSO, and therefore the GSO data were not utilized.

4.2 Data Transmission and Data Processing on Ground

Observation data of the XIS and HXD are stored on the onboard data recorder, together with the satellite housekeeping data such as attitude or temperature of subsystems, and then transmitted via downlink to the ground station at Unichinoura, Kagoshima, Japan, while the satellite passes over it. Out of 15 daily revolutions of *Suzaku* around the Earth, such a "ground contact" takes place 4–5 times, each with typical duration of 10–15 min per pass. The size of data which can be transmitted to ground in a day is roughly 500–600 MB.

Since the XIS and the HXD are photon-counting instruments, event lists which contain every single X-ray photon (and background signal) are produced as raw products, and saved as so-called FITS files, together with the satellite housekeeping information. FITS, or Flexible Image Transport System, is the standard file format for astronomical data maintained by NASA. To extract X-ray spectra and light curves of individual celestial sources from the FITS files, we utilized the analysis software package HEASOFT, or High Energy Astrophysics Software, of version 6.9 together with a detector calibration database which was latest as of 2010 April. HEASOFT is the standard analysis tools distributed and maintained mainly by NASA/Goddard Space Flight Center, and available from the internet[1]. Spectral analyses were done using the XSPEC spectral fitting package version 12 contained in HEASOFT.

[1] http://heasarc.gsfc.nasa.gov/docs/software/lheasoft/

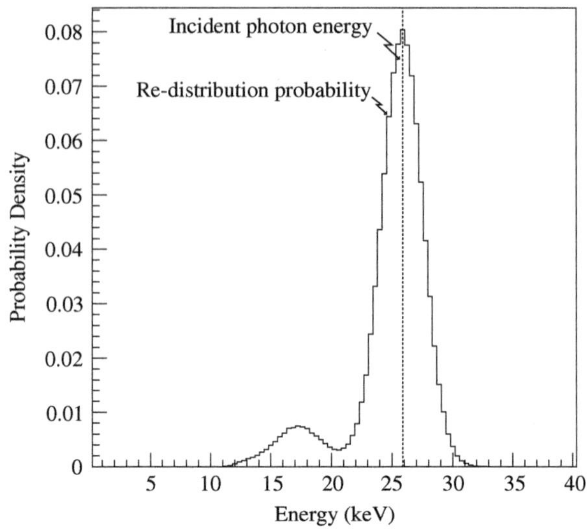

Fig. 4.13 The HXD/PIN energy re-distribution function for an incident X-ray photon with an initial energy of 25.7 keV. The ordinate represents probability density for each photon to be recorded into an energy bin in the abscissa

4.3 Model Fit to an Observed Spectrum

When extracting physical information from observed spectra, we usually fit them with spectral models. A model predicts a flux of X-ray photons (photon rate) emitted from a celestial X-ray source, and by changing parameters, we try to find a set of parameters which nicely reproduces an observed spectral features. However, practically, an energy response of an X-ray detector is generally complex, and the energy of an incoming photon is not necessarily recorded intact, but usually with slight modification by the detector response, or being re-distributed over energies.

For example, as presented in Fig. 4.13, the HXD/PIN has a response which is broadened and distorted from an ideal (delta-function like) one due to electrical read-out noise and Compton scattering happening in the Si diodes. To disentangle (or deconvolve) the effect of the effective area and the response from observed data is an inverse problem, and difficult unless we have a true spectral shape which we would like to estimate from the observed data.

Therefore, we perform a fitting not in the photon rate space but in the count rate space where the model is convolved with the effective area and the detector response. Fitting a spectrum with a model reduces to minimize deviations $\delta(E_i)$ defined for energy bins E_i as

$$\delta(E_i) = |D(E_i) - \Sigma_j R(E_j, E_i) \times M(E_j)|,$$

where $D(E_i)$, $R(E_j, E_i)$, and $M(E_j)$ are an observed count rate in an energy bin E_i, a detector response function (or probability of re-distribution from E_j to E_i), and a photon flux predicted by a model, respectively.

To see the goodness of a fit, we use the χ^2 statistics. In this, minimization is performed for χ^2 which is defined as

$$\chi^2 = \Sigma_i \left(\frac{\delta(E_i)}{\Delta D(E_i)} \right)^2,$$

where $\Delta D(E_i)$ is an error associating the data in the energy bin. Usually, we estimate $\Delta D(E_i)$ using the statistical error $\sqrt{N}$ of the raw count and the Gaussian error propagation when subtracting the background. The value of χ^2 expresses how largely the data and the model differ over the statistical fluctuation. If the model reproduces the data nicely, we expect that $\chi^2 \approx \nu$, where ν is the number of degree of freedom in the fit because each bin, or each element inside the summation, should statistically fluctuate around 0 with a typical width of 1σ.

When minimization of χ^2 converges, we calculate the null hypothesis probability. Our hypothesis here is that the data is a sample extracted from a general population which is described by the model. If we get very low probability, the hypothesis is statistically rejected. On the other hand, moderately high probability means that the data *could be* derived from the general population (the model). Note, however, that even if the resulting null hypothesis probability is high, this does not ensure correctness of the model. The fitting statistics can be used just as an indicator of similarity between the spectral shapes of the data and the model, and physical arguments other than fittings (e.g., Chap. 3 and Sect. 5.3) should justify the model.

References

Fukazawa, Y., Mizuno, T., Watanabe, S., et al. 2009, PASJ, 61, 17
Kokubun, M., Makishima, K., Takahashi, T., et al. 2007, PASJ, 59, 53
Koyama, K., Tsunemi, H., Dotani, T., et al. 2007, PASJ, 59, 23
Mitsuda, K., Bautz, M., Inoue, H., et al. 2007, PASJ, 59, 1
Nakajima, H., Yamaguchi, H., Matsumoto, H., et al. 2008, PASJ, 60, 1
Serlemitsos, P. J., Soong, Y., Chan, K.-W., et al. 2007, PASJ, 59, 9
Takahashi, T., Abe, K., Endo, M., et al. 2007, PASJ, 59, 35
Tawa, N., Hayashida, K., Nagai, M., et al. 2008, PASJ, 60, 11
Uchiyama, H., Ozawa, M., Matsumoto, H., et al. 2009, PASJ, 61, 9

Chapter 5
Estimating the Masses of White Dwarfs in Magnetic Cataclysmic Variables

In this chapter, we apply the X-ray spectral emission model of magnetic cataclysmic variables (CVs) constructed in Chap. 3 to wide-band X-ray data obtained by *Suzaku* observations. In Sect. 5.1, we describe our target selection criteria and data processing method. Section 5.2 explains method of the spectral fitting and the derived WD mass estimations. Systematic uncertainties which could be involved in the present analysis are discussed in Sect. 5.3, and then our results are compared with previous studies in Sect. 5.4.

5.1 Observation and Preparation of Data Set

5.1.1 Target Selection and Involved Bias

A sample of WD binaries was constructed by assembling 17 Intermediate Polar (IP) binaries observed with *Suzaku*. Table 5.1 lists information of their *Suzaku* observations, together with their background-subtracted count rates recorded with the XIS and HXD/PIN. Seven of the 17 observations were performed according to successful proposals submitted by our group (see PI column in Table 5.1). As mentioned in Sect. 3.2, we concentrate on IPs among magnetic CVs, because estimating WD masses in Polars involves additional difficulty in modeling effects of their stronger magnetic fields, and IPs dominate the GRXE in hard X-ray energy band ($E > 10\,\mathrm{keV}$).

Since the sample has been constructed by combining different observations, it is heterogeneous and incomplete. However, when compared with, for example, magnetic CVs detected by the hard X-ray all-sky survey by *Swift*/BAT (Brunschweiger et al. 2009), it can be approximately regarded as a flux-limited sample, or a subset of it more strictly; the present sample covers $\sim$80 % of the IPs detected by the all-sky survey. To perform a detailed spectral analysis, we selected targets with enough statistics in the hard X-ray band; only sources with background-corrected

T. Yuasa, Suzaku *Studies of White Dwarf Stars and the Galactic X-ray Background Emission*, Springer Theses, DOI: 10.1007/978-4-431-54219-3_5, © Springer Japan 2013

Table 5.1 *Suzaku* observations of 17 IPs analyzed in the present study

System	Coordinate[a]		Start time	Exp.[b]	Energy band[c]		Count rate[d]		Obs. ID	P.I.[e]
	RA	Dec	UT	ks	XIS (keV)	HXD (keV)	XIS	HXD		
FO Aquarii	334.481	−8.351	2009-06-05 08:14	33.4	3.0–12.0	17.0–40.0	0.95	0.08	404032010	T. Yuasa
XY Arietis	44.036	19.442	2006-02-03 23:02	93.6	3.0–12.0	12.0–50.0	0.65	0.07	500015010	*Suzaku* Team
MU Camelopadalis	96.318	73.578	2008-04-14 00:55	50.1	3.0–12.0	12.0–40.0	0.27	0.06	403004010	M. Ishida
BG Canis Minoris	112.871	9.94	2009-04-11 12:11	45.0	3.0–12.0	12.0–40.0	0.51	0.10	404029010	T. Yuasa
V709 Cassiopeiae	7.204	59.289	2008-06-20 10:24	33.3	3.0–12.0	12.0–50.0	1.10	0.16	403025010	K. Ebisawa
TV Columbae	82.356	−32.818	2008-04-17 18:00	30.1	3.0–12.0	12.0–50.0	1.21	0.20	403023010	T. Yuasa
TX Columbae	85.834	−41.032	2009-05-12 16:19	51.1	3.0–9.0	12.0–40.0	0.29	0.04	404031010	T. Yuasa
YY Draconis	175.896	71.703	2008-06-15 18:37	27.4	3.0–12.0	12.0–40.0	0.85	0.10	403022010	T. Yuasa
PQ Geminorum	117.822	14.74	2009-04-12 13:46	43.2	3.0–10.0	13.0–50.0	0.56	0.08	404030010	T. Yuasa
EX Hydrae	193.107	−29.249	2007-07-18 21:23	91.0	3.0–12.0	12.0–33.0	2.12	0.14	402001010	M. Ishida
NY Lupi	237.061	−45.478	2007-02-01 15:17	86.8	4.0–12.0	13.0–50.0	0.62	0.16	401037010	K. Mukai
V2400 Ophiuchi	258.146	−24.247	2009-02-27 11:42	110	3.0–12.0	12.0–50.0	2.60	0.21	403021010	H. Mori
AO Piscium	343.825	−3.178	2009-06-22 11:50	35.6	3.0–12.0	17.0–50.0	1.21	0.05	404033010	T. Yuasa
V1223 Saggitarii	283.759	−31.163	2007-04-13 11:31	46.2	4.0–12.0	12.0–50.0	2.38	0.48	402002010	M. Ishida
RX J2133.7+5107	323.432	51.124	2006-04-29 06:50	62.8	4.5–12.0	12.0–50.0	0.55	0.15	401038010	K. Mukai
IGR J17303−0601	262.59	−5.993	2009-02-16 10:09	27.7	3.0–12.0	13.0–50.0	0.56	0.16	403026010	K. Ebisawa
IGR J17195−4100	259.898	−41.015	2009-02-18 11:03	26.9	3.0–12.0	12.0–50.0	1.02	0.15	403028010	K. Ebisawa

[a] In units of degree (J 2000 equinox)
[b] Net exposure
[c] Energy band used in the spectral analysis (Sect. 5.2.2)
[d] Count rates in units of counts s^{-1} integrated over the noted energy bands. Both of the NXB and the CXB are subtracted
[e] Name of the principal investigator, or proposer, of the observation

HXD count rates higher than 0.04 counts s^{-1} are used in the present analysis. This rate is equivalent to about 10 % of the typical HXD background count rate (Kokubun et al. 2007; Fukazawa et al. 2009). With these criteria, some IPs (AE Aqr, GK Per, RXS J070407.9+26250, and RXS J180340.0+40121) were excluded from the sample, although *Suzaku* has observed them. By converting the count rate, the limiting hard X-ray flux for the present sample becomes about 1 mCrab, corresponding to an energy flux of 1.6×10^{-8}erg cm^{-2} s^{-1} in the 12–40 keV band.

5.1.2 Data Processing and Extraction of Spectra

The typical energy bands utilized in the present study are 3–12 keV and 12–50 keV for the XIS and the HXD, respectively (Table 5.1). Due to lower signal-to-noise ratio of the XIS BI spectrum in higher energies, we simply discarded the spectrum above 8 keV. In some observations, especially in that of FO Aqr, we raised the HXD lower energy bound since an enhanced electrical noise contaminated the data therein. The HXD data above 40 keV are discarded when the source signals were below 5 % of the background (the 1σ systematic errors of background estimation is 1 %; see Sect. 4.1.2).

We excluded the XIS and HXD/PIN data taken when the target is occulted by the Earth, or the spacecraft was in the South Atlantic Anomaly where observations are stopped due to the extremely high particle background. To reduce the contamination from scattered Solar X-rays, we also removed the XIS data taken while the target elevation angle above the Sun-lit Earth rim was less than 20°. We applied the same event filtering criteria as used for creating nominal cleaned event files, as described at the *Suzaku* website.[1]

We present an example of X-ray images taken with the XIS in Fig. 5.1. In the XIS data analysis, we extracted time-averaged spectra of the targets using circular regions with radii of 2′ around the brightness peak in the obtained CCD images. A larger radius would increase the instrumental and celestial backgrounds and reduce the signal-to-noise ratio. Background spectra were also produced from the same observation data, excluding apparent point sources if any. By using this background spectrum, we can simultaneously subtract non-X-ray instrumental background (NXB) and the cosmic X-ray background (CXB) from the observed data. Because the signal-to-noise ratio is sufficiently high for all targets, the vignetting effect of the XRT (Serlemitsos et al. 2007) can be neglected in the background subtracting procedure. We generated redistribution matrices and auxiliary response files of the XIS/XRT using `xisrmfgen` and `xissimarfgen` (Ishisaki et al. 2007), respectively (both contained in HEASOFT). In the spectral analysis, we summed the spectra taken by the FI sensors, and used a detector response averaged over them.

We similarly accumulated a time-averaged HXD/PIN spectrum of each target, and subtracted the non-X-ray instrumental background extracted from the NXB event file

[1] http://www.astro.isas.jaxa.jp/suzaku/process/v2changes/criteria_xis.html

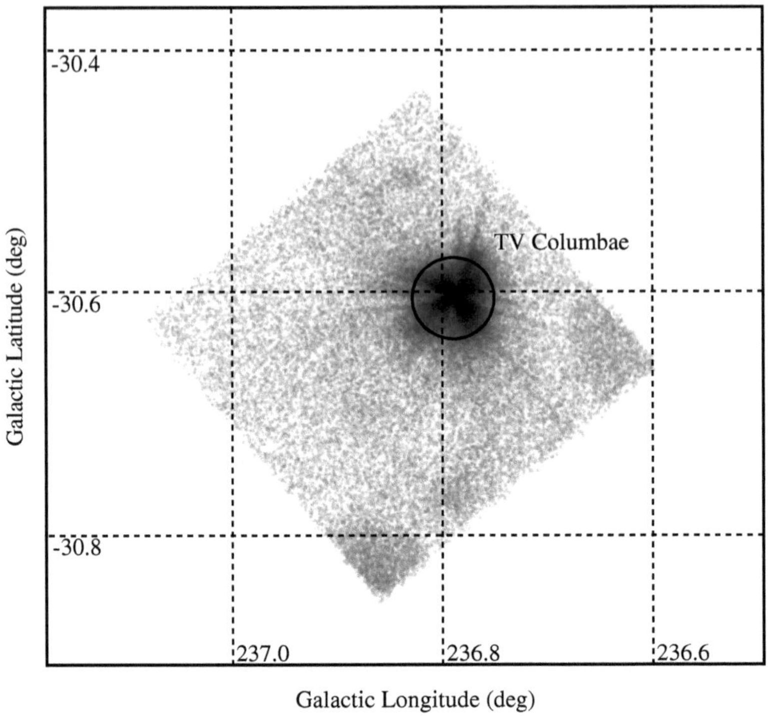

Fig. 5.1 An example of X-ray image of TV Columbae observed by the XIS. Due to the point spread function of the XRT (Sect. 4.1.1), the image is not like a *point* but spreads like a cross. *Solid circle* represents the region used to extract spectrum

distributed via the HXD team (Sect. 4.1.2). The CXB component were simulated after a model of Boldt (1987), i.e., a surface brightness defined as

$$\mathrm{CXB}(E) = 9.0\times 10^{-9}(E/3\,\mathrm{keV})^{-0.29}\exp(-E/40\,\mathrm{keV})\,\mathrm{erg\,cm^{-2}\,s^{-1}\,sr^{-1}}, \quad (5.1)$$

and then subtracted from the data as well. When fitting an HXD/PIN spectrum, we employed an appropriate redistribution matrix file contained in CALDB, depending on the pointing mode and the observation date (Sect. 4.1.2). Because the observation of XY Ari was performed at an off-axis position of the HXD, the effective area of HXD/PIN decreased by $\sim$20 %; we took this effect into account by applying an additional correction to the effective area using a file calculated with `hxdarfgen`.

Figures 5.2 and 5.3 show examples of thus produced wide-band spectra of selected targets. Because we selected observations with significances better than the accuracy of the NXB subtraction, the celestial signals (i.e., X-ray photons from IPs) are clearly detected by both instruments.

In some observations, the XIS images contained weak point sources other than the target IPs. These sources could contaminate the data of HXD/PIN because it has

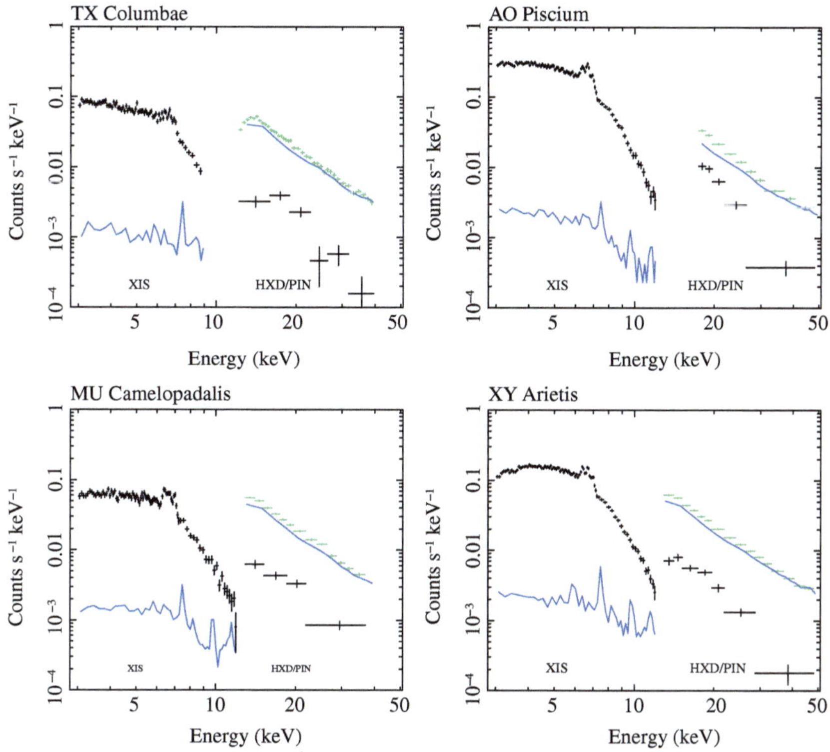

Fig. 5.2 Wide-band X-ray spectra of four objects which exhibited low count rates in the HXD/PIN energy band. *Green crosses* and *blue solid curves* are raw signal counts and the NXB, respectively. *Black crosses* are signals from the target IPs derived by subtracting the NXB and the CXB from the raw data. The raw XIS data are almost the same as the *black crosses*, and hence not displayed for better visibility

a wider field of view ($34' \times 34'$ FWHM; Sect. 4.1.2) and no imaging capability. However, their contributions in the HXD/PIN energy band are estimated to be $\lesssim$2–3 % of background-subtracted signals in the particular case of XY Ari, and much lower in other observations (i.e., target IPs are much brighter than the other neighbor sources). Therefore, we simply neglected them in the present analysis.

5.2 Deriving WD Masses by Fitting X-ray Spectra

5.2.1 Construction of a Fit Model

To estimate the masses of WDs from the observed spectra, we have to fit them with an appropriate spectral model which correctly accounts for emission and intervening

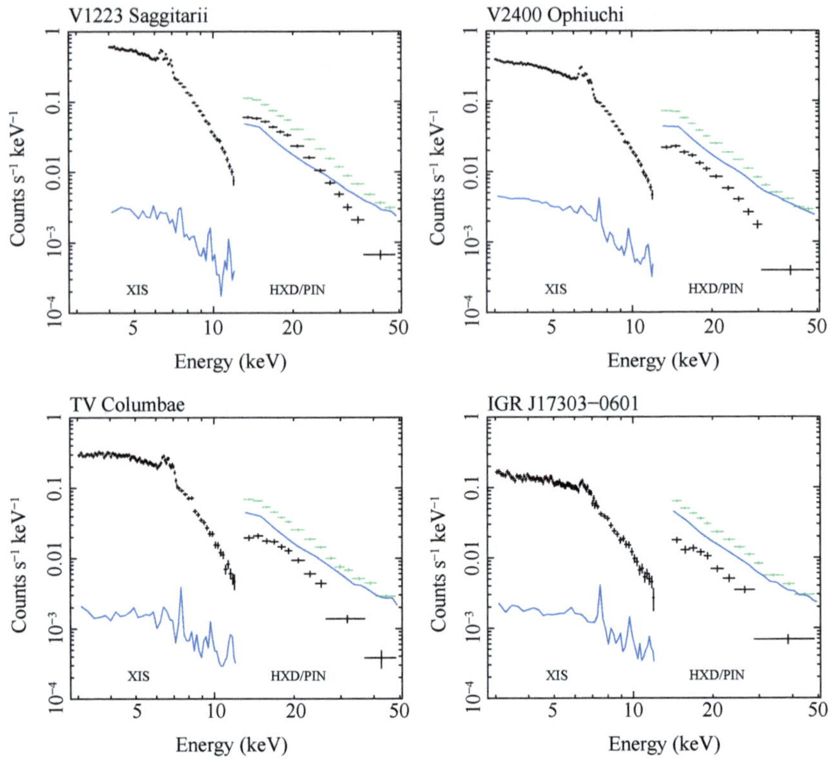

Fig. 5.3 Wide-band X-ray spectra of four objects which exhibited high count rates in the HXD/PIN band. See Fig. 5.2 for description of the plots

absorbing material. Although an elaborate numerical model (the IP model) has already been constructed in Chap. 3, below we start from a simple model, and fit the spectra of XIS-FI, XIS-BI, and HXD/PIN simultaneously using it. To take into account a cross-instrument renormalization factor which depends on the satellite attitude, a model for the HXD/PIN spectrum was rescaled by a factor of 1.16 (XIS-aimed observations) or 1.18 (HXD-aimed observations); see *Suzaku* Memo 2008-06 for details. Initially, we use the energy bands 1–12 and 12–50 keV for the XIS and HXD/PIN spectra. Hereafter, the quoted errors are at 90 % confidence levels unless otherwise noted.

5.2.1.1 Failing Single-Temperature CIE Plasma Model

As a first attempt to quantify the observed spectra, we fitted them with a very simple model which consists of a single-temperature collisional ionization equilibrium (CIE) plasma emission subjected to the interstellar absorption with a hydrogen column density n_{H}, and emission lines from neutral iron (Fe I Kα and β at 6.4 and 7.1 keV).

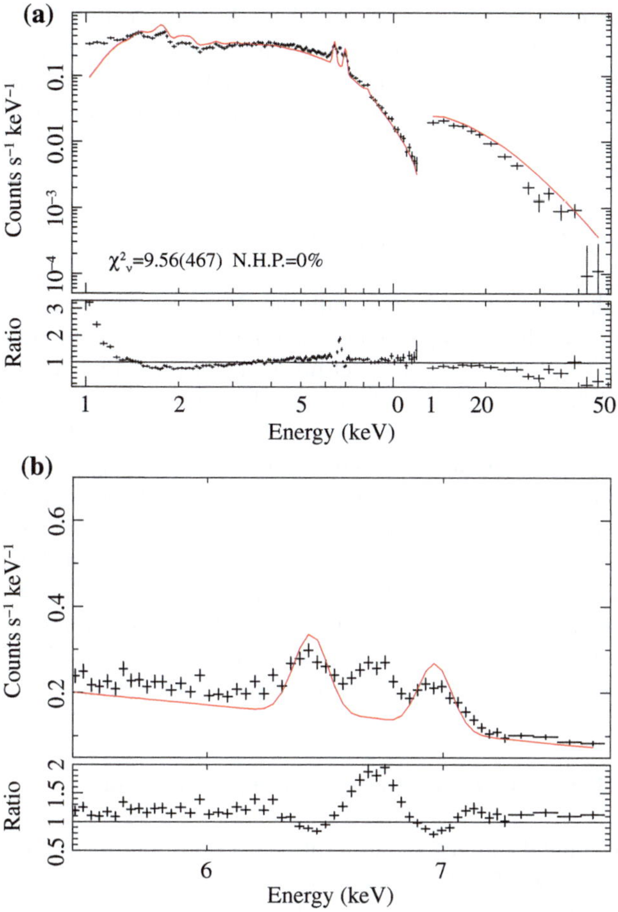

Fig. 5.4 **a** An example of the fit with a single-temperature CIE plasma model. Observed spectrum of TV Col (*crosses*) and the best-fit model (*red curve*) are plotted in the *upper panel*. *Lower panel* shows ratios of the spectrum to the model. **b** A close-up of the same spectrum around the Fe line region

These neutral emission lines are thought to arise from reprocessing (reflection) of hard X-rays at cool gas, but we just model them using the Gaussians because detailed modeling is difficult, and they do not affect mass estimation result (see Sect. 5.3.4). Figure 5.4 shows an example of a fit to the TV Col spectrum. The model obviously fails to reproduce the observed spectrum especially in the lower energies 1–3 keV, and at the Fe XXV (He-like) Kα emission line. Although we concentrate on the TV Col spectrum in this section for clarity, similar results (failing fits) are obtained in the other targets. This implies that the observed continuum shape and the Fe line structures cannot be explained simultaneously by the single-temperature model. In this TV Col case, the model converged at a high temperature of 64 keV to reproduce the hard continuum, leaving only the Fe XXVI (H-like) Kα line. Such a hot plasma does not emit Fe XXV Kα line photons because Fe ions are almost fully ionized at this temperature. Such a failure of the single-temperature model was already expected in Chap. 3, because X-ray spectra of IPs should have the multi-temperature nature.

5.2.1.2 Failing Single-Absorption IP Model

We replaced the CIE plasma model with the IP model constructed in Chap. 3, in an attempt to solve the discrepancy between the conflicting continuum and Fe line requirements. The same TV Col spectrum is again presented with the best-fit IP model in Fig. 5.5a. Although the deficiency of Fe XXV Kα line was explained away by the lower temperature plasma component at the lower part of the PSR (Fig. 3.5), the model still poorly reproduces the continuum shape, leaving $\chi^2/\nu = 14.14(457)$.

While the Fe-K line structure became much better explained, significant fit residuals in the lower energy band (<2 keV) suggest inappropriate modeling of the photo absorption. As can be seen in the geometry of IPs (Fig. 2.14), an IP is thought to be fed with an accretion gas stream arising from a truncated accretion disk. As the WD spins and orbits around the companion star, the PSR is sometimes shrouded by the accreting (pre-shock) gas as illustrated in Fig. 5.6, and the softer (i.e., lower energy) end of the PSR emission will decrease due to photo absorption.

5.2.1.3 The IP Model with Partially-Covering Absorption

To take into account the effects of the additional absorption, we introduced a concept of partially-covering (PC) photo absorption. This assumes that some fraction of the incident emission (i.e., X-rays from the PSR) suffers from an absorption with a hydrogen column density $n_{\rm H,PC}$, and the remaining will directly reach us. This is an approximation to the time-variable absorption (Fig. 5.6) and the non uniformity of the gas density in the accretion stream itself. The total model thus becomes,

$$
\begin{aligned}
&\texttt{Total Model}(\text{photons s}^{-1}\,\text{cm}^{-2}) \qquad (5.2)\\
&= \texttt{Constant} \times [\texttt{Interstellar Photo Absorption}(n_{\rm H})]\\
&\quad \times (C_{\rm PC}\cdot[\texttt{Photo Absorption}(n_{\rm H,PC})]\cdot \texttt{IP\_PSR} + (1-C_{\rm PC})\cdot \texttt{IP\_PSR}\\
&\quad + \texttt{Gaussian} + \texttt{Gaussian}).
\end{aligned}
$$

The `Constant` is a constant factor used to adjust the cross-instrument normalizations (see above). The terms with $C_{\rm PC}$ and $(1 - C_{\rm PC})$ represent the absorbed and non-absorbed components, respectively. These two `IP_PSR` components share the same parameters ($M_{\rm WD}$, $Z_{\rm Fe}$, and normalization). The two Gaussians are for the Kα and Kβ emission lines from neutral irons.

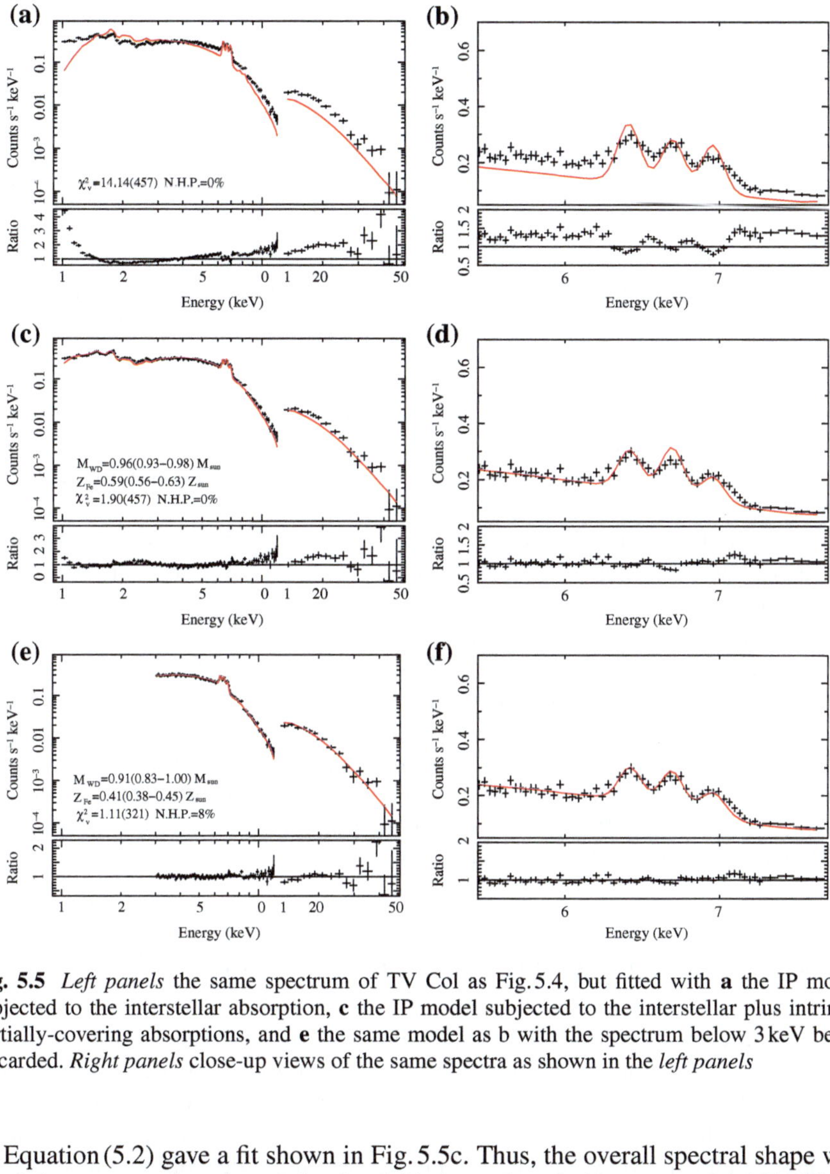

Fig. 5.5 *Left panels* the same spectrum of TV Col as Fig. 5.4, but fitted with **a** the IP model subjected to the interstellar absorption, **c** the IP model subjected to the interstellar plus intrinsic partially-covering absorptions, and **e** the same model as b with the spectrum below 3 keV being discarded. *Right panels* close-up views of the same spectra as shown in the *left panels*

Equation (5.2) gave a fit shown in Fig. 5.5c. Thus, the overall spectral shape was much better reproduced, and the fit yielded a reduced chi-square of 1.90(457). A WD mass of $0.96^{+0.02}_{-0.03}\,M_{\odot}$ was obtained. Although the value significantly improved from that of the single-absorption IP model, it is not yet formally acceptable, and residuals are still recognizable especially in the lower energy band and energies above 10 keV. We consider that the lower energy residuals are caused by still insufficient modeling of absorption. According to previous studies of magnetic CVs, it is known that intrinsic absorption has a complex structure which should be approximated by an absorption column density with a continuous gradient (e.g., Done et al. 1995;

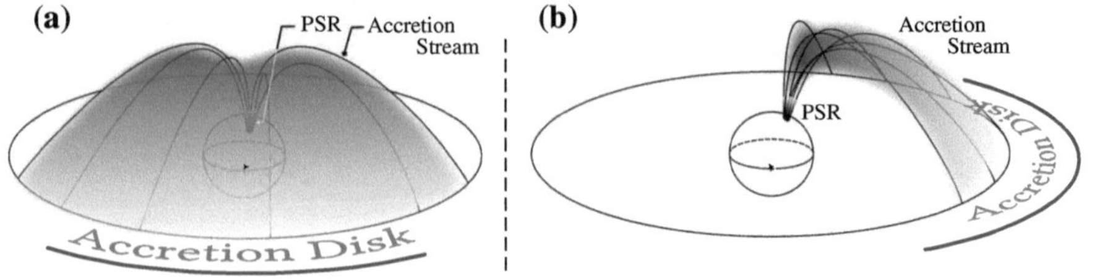

Fig. 5.6 A schematic geometry of the complex intrinsic absorption in an IP. The observer's line of sight is perpendicular to the paper. **a** X-rays from the PSR suffer from absorption by the accretion stream. **b** The PSR is directly observed, without obscured by the accretion stream

Done and Magdziarz 1998) unlike discrete value as used above. In addition, the spin- and orbit-dependent absorptions result effectively in a "distributed" absorption when time averaged. The PC formalism gives only an approximation to this type of complex absorption.

Ezuka and Ishida (1999) showed that the PC formalism recovers its validity if the spectra of lower energies (e.g., <3 keV) are discarded. Therefore, we followed their prescription, and ignored our XIS spectrum below 3 keV. Then, as shown in Fig. 5.5e, the fit statistics were further improved to $\chi^2_\nu = 1.11(321)$ which is now formally acceptable with a null hypothesis probability of 8 %. The continuum spectral shape and the Fe emission lines were better reproduced compared to the previous fit (panel b of Fig. 5.5). The best-fit $M_{\rm WD}$ of $0.91^{+0.09}_{-0.08}$ $M_\odot$ remained unchanged, within errors.

Based on the fit goodness, we consider that this partially-covered IP model applied to the spectra above 3 keV derives reasonable WD masses. In the next section, we use this model to estimate WD masses also in the remaining targets as well. Hereafter, the XIS data are used only in energies above ~3 keV.

5.2.2 WD Masses Estimated from Wide-Band Spectral Fitting

We fitted the spectra of the 17 IPs, including already analyzed TV Col, using the IP model suffered from the partially-covering absorption. Two targets, FO Aqr and V2400 Oph, required us to introduce an additional absorbed component (i.e., two absorption column densities in total), because the fits were unacceptable without it. Figures 5.7 and A.1 show the spectra and the best-fit models of individual IPs. As evidenced by the best-fit parameters listed in Table 5.2, the model successfully reproduced the observed spectra in most cases except for that of EX Hya (see below). Looking at the spectra, for example those presented in Fig. 5.7, it can be recognized that the HXD/PIN spectrum of FO Aqr is more strongly curved than that of XY Ari. This feature indicates that FO Aqr has a lower PSR shock temperature and hence lower WD mass than XY Ari; derived masses, $0.51^{+0.09}_{-0.04}$ $M_\odot$ and $0.95^{+0.10}_{-0.12}$ $M_\odot$ for FO Aqr and XY Ari, respectively, are consistent with this apparent properties. The table also gives the shock temperatures which are related to $M_{\rm WD}$ through Fig. 3.4. The derived

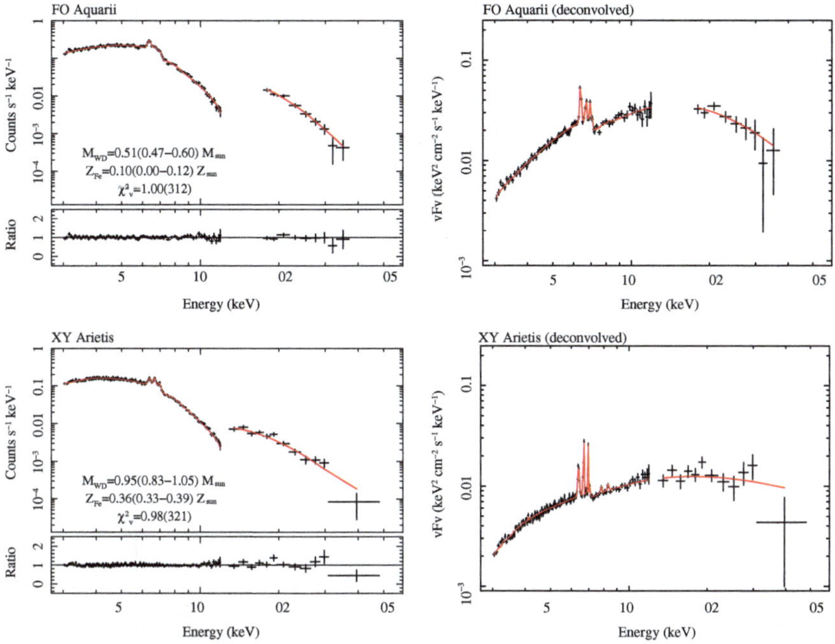

Fig. 5.7 Observed XIS and HXD spectra (*black crosses*) of two representative IPs in our sample compared with the best-fit model predictions (*solid curves*). Below 12 keV, only the spectra extracted from the XIS-FI are presented, and those from the XIS-BI are omitted for clarity although they were actually used in the fittings. *Left panels* raw spectra, wherein the detector responses are not removed. The *lower panels* show the data-to-model ratio. *Right panels* the same spectra, shown after removing the detector responses and multiplied by E^2, or equivalent to the νF_ν representation. *Fits to the remaining objects are shown in Fig. A.1 of Appendix A*

masses are in the range of 0.4–1.3 $M_\odot$ which is consistent with current understanding on the WD mass and its upper limit of $\sim$1.4 $M_\odot$ (Sect. 2.2.1).

Some sources, NY Lup, V1223 Sgr, and RX J2133 showed worse fitting statistics (larger χ^2) when their XIS spectra were used from 3 keV, instead of the values we finally set (4 or 4.5 keV; Table 5.1). In these cases, significant residuals were seen only in the 3–3.5 keV band; this might represent a softer blackbody emission originating from the WD surface heated by the hard X-rays, or a soft thin-thermal emission (e.g., Haberl et al. 2002). Since such a soft component does not affect our WD mass estimations, we simply used the increased lower energy threshold for the XIS spectra of these objects.

As shown in Table 5.2, the fit for EX Hya was rather poor, $\chi^2/\nu = 414/316$, with a corresponding null hypothesis probability of 1.7×10^{-4}. This is caused by residuals in the 6.7–7.0 keV band, where the modeled spectral shape of the Fe Kα lines slightly differs from that of the data as presented in Fig. 5.8. The complex fine structure of the Fe lines seen in a high resolution spectrum of this source obtained with the *Chandra* HETG (Hellier and Mukai 2004) probably manifests a cause of

Table 5.2 The best-fit spectral parameters obtained with the partial-covering absorber model. The errors are at 90 % confidence levels

System	Z_{Fe} ($Z_\odot$)	M_{WD} ($M_\odot$)	kT_s (keV)	nH^a 10^{22} cm^{-2}	nH^b_{PC} 10^{22} cm^{-2}	C^c_{PC}	$\chi^2_\nu(\nu)$	$F^d_{2,10}$	$F^e_{12,40}$
FO Aqr	$0.10^{+0.02}_{-0.10}$	$0.51^{+0.09}_{-0.04}$	$17.0^{+4.9}_{-2.0}$	$9.54^{+1.38}_{-2.26}$	$398.0^{+204.0}_{-93.5}$	$0.68^{+0.10}_{-0.24}$	1.00(312)	3.73	1.69
					$41.5^{+7.8}_{-8.2}$	1.0 (fixed)			
XY Ari	$0.36^{+0.03}_{-0.03}$	$0.95^{+0.10}_{-0.12}$	$49.9^{+12.4}_{-11.7}$	$9.62^{+0.43}_{-0.36}$	$115.0^{+33.1}_{-19.9}$	$0.42^{+0.06}_{-0.02}$	0.98(321)	1.42	0.58
MU Cam	$0.49^{+0.07}_{-0.06}$	$0.95^{+0.21}_{-0.22}$	$49.9^{+30.8}_{-19.5}$	$5.42^{+1.34}_{-0.98}$	$92.9^{+39.6}_{-22.6}$	$0.57^{+0.06}_{-0.04}$	1.12(318)	0.92	0.37
BG CMi	$0.18^{+0.04}_{-0.03}$	$1.04^{+0.12}_{-0.10}$	$61.0^{+19.8}_{-12.1}$	$6.83^{+1.64}_{-1.93}$	$40.2^{+18.6}_{-11.2}$	$0.46^{+0.10}_{-0.08}$	1.13(318)	2.01	0.71
V709 Cas	$0.16^{+0.04}_{-0.03}$	$1.07^{+0.10}_{-0.15}$	$65.2^{+17.6}_{-18.5}$	$2.83^{+0.48}_{-0.45}$	$101.0^{+22.9}_{-14.2}$	$0.46^{+0.03}_{-0.02}$	1.05(321)	3.63	1.18
TV Col	$0.41^{+0.04}_{-0.03}$	$0.91^{+0.09}_{-0.08}$	$45.7^{+10.0}_{-7.4}$	$4.48^{+0.88}_{-1.06}$	$45.8^{+16.9}_{-12.1}$	$0.39^{+0.05}_{-0.05}$	1.11(321)	4.67	1.44
TX Col	$0.41^{+0.09}_{-0.09}$	$0.80^{+0.27}_{-0.15}$	$35.8^{+29.5}_{-10.8}$	$1.75^{+1.64}_{-1.75}$	$32.4^{+22.4}_{-14.7}$	$0.41^{+0.14}_{-0.13}$	1.23(186)	1.24	1.12
YY Dra	$0.59^{+0.07}_{-0.06}$	$0.93^{+0.15}_{-0.15}$	$47.8^{+19.0}_{-13.6}$	$0.00^{+0.69}_{-0.00}$	$37.5^{+51.7}_{-25.8}$	$0.18^{+0.08}_{-0.08}$	0.98(318)	3.37	0.70
PQ Gem	$0.17^{+0.04}_{-0.03}$	$1.09^{+0.09}_{-0.13}$	$68.3^{+16.6}_{-17.3}$	$2.91^{+1.25}_{-2.09}$	$46.6^{+21.0}_{-16.6}$	$0.43^{+0.10}_{-0.06}$	1.14(303)	2.21	0.66
EX Hya[g]	$0.55^{+0.04}_{-0.03}$	$0.43^{+0.02}_{-0.02}$	$13.1^{+0.9}_{-0.9}$	$0.81^{+0.23}_{-0.19}$	$99.6^{+13.5}_{-10.6}$	$0.42^{+0.04}_{-0.05}$	1.31(316)	8.49	1.32
NY Lup	$0.42^{+0.03}_{-0.02}$	$1.26^{+0.04}_{-0.03}$	$106.1^{+14.9}_{-9.0}$	$8.60^{+0.88}_{-0.96}$	$202.0^{+35.3}_{-28.4}$	$0.55^{+0.04}_{-0.03}$	1.16(251)	2.80	1.22
V2400 Oph	$0.25^{+0.03}_{-0.03}$	$0.66^{+0.05}_{-0.06}$	$25.6^{+3.4}_{-3.7}$	$1.93^{+0.75}_{-1.20}$	$302.0^{+65.5}_{-46.9}$	$0.58^{+0.07}_{-0.09}$	0.92(319)	4.61	1.55
					$38.4^{+12.8}_{-11.9}$	1.0 (fixed)			
AO Psc	$0.46^{+0.08}_{-0.04}$	$0.63^{+0.07}_{-0.03}$	$23.7^{+4.6}_{-1.8}$	$2.37^{+1.53}_{-2.37}$	$24.9^{+8.4}_{-8.4}$	$0.51^{+0.16}_{-0.12}$	1.18(317)	4.31	1.04
V1223 Sgr	$0.27^{+0.01}_{-0.02}$	$0.79^{+0.04}_{-0.04}$	$35.0^{+3.3}_{-3.1}$	$9.91^{+0.65}_{-0.61}$	$201.0^{+25.9}_{-9.9}$	$0.50^{+0.03}_{-0.02}$	1.12(252)	8.29	3.49
RX J2133	$0.30^{+0.04}_{-0.03}$	$0.83^{+0.13}_{-0.10}$	$38.3^{+12.8}_{-7.9}$	$12.50^{+1.47}_{-1.76}$	$330.0^{+64.6}_{-50.1}$	$0.62^{+0.05}_{-0.04}$	1.22(194)	2.04	4.57
IGR J17303	$0.23^{+0.04}_{-0.04}$	$1.24^{+0.06}_{-0.39}$	$100.0^{+21.1}_{-59.9}$	$5.21^{+0.49}_{-0.46}$	$278.0^{+34.5}_{-27.2}$	$0.74^{+0.03}_{-0.02}$	1.19(320)	1.79	1.01
IGR J17195	$0.32^{+0.04}_{-0.04}$	$0.94^{+0.11}_{-0.15}$	$48.9^{+13.5}_{-13.9}$	$2.98^{+0.55}_{-0.54}$	$101.0^{+36.8}_{-10.7}$	$0.42^{+0.07}_{-0.04}$	0.95(321)	3.48	1.09

[a] The absorption column density of fully-covering absorbent
[b] The absorption column density of the partial-covering absorbent (see text)
[c] The partial-covering fraction
[d] 2–10 keV flux in units of 10^{-11} erg cm^{-2} s^{-1}
[e] 12–40 keV flux in units of 10^{-11} erg cm^{-2} s^{-1}
[f] The column density of the additional absorbed component (see text)

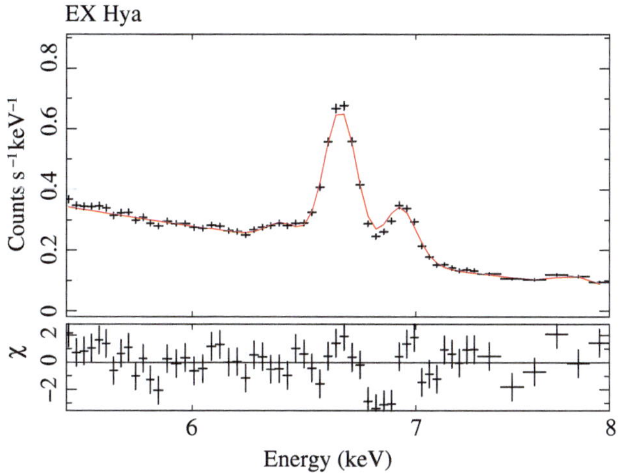

Fig. 5.8 *Upper panel* a close-up view of the XIS-FI spectrum of EX Hya. *Red curve* is the best-fit IP model which leaves residuals in the 6.7–7.0 keV band. *Lower panel* fit residuals expressed in terms of χ (Sect. 4.3)

this problem: this source exhibits a strong resonance line and dielectric satellite lines in the 6.6–6.7 keV band, and this makes the target somewhat peculiar among IPs. The worse fit is probably explained by (1) this microscopic line structure which is not correctly modeled by the IP model, and (2) a fact that our EX Hya spectrum has high counting statistics compared to those of the other targets. In the present analysis, we just use the best-fit model including the Fe lines (Table 5.2) to derive the WD mass in EX Hya since further studies on the fine structure of the Fe lines are not possible with the energy resolution of *Suzaku*. It should be noted that even when the Fe line complex was ignored in the fitting, the estimated WD masses were the same (0.43 $M_{\odot}$) within fitting errors of ± 0.02 $M_{\odot}$, and the fitting statistics improves to an acceptable level (null hypothesis probability >1 %). This can be understood as a result of strong constraint on the shock temperature imposed by the HXD spectral shape which shows a clear thermal cutoff (see EX Hya spectrum in Fig. A.1).

After considering the several exceptional cases, we obtained the average WD mass of 0.88 ± 0.23 $M_{\odot}$ and the mean iron abundance 0.33 ± 0.14 $Z_{\odot}$ from the present sample including 17 IPs (associated errors are 1σ standard deviations for the distributions).

5.2.3 Comparison of Fittings Between the XIS and HXD/PIN Energy Bands

Compared to the previous studies (Sect. 2.2.5), one of the evident advantage of the present study is that the information on the PSR temperature is obtained from two independent spectral features; the Fe emission lines and the hard X-ray continuum.

Table 5.3 WD masses derived from the PIN-only and the XIS-only fittings

System	$M_{\rm WD}^{\rm PIN}$ ($M_\odot$)	$M_{\rm WD}^{\rm XIS}$ ($M_\odot$)	System	$M_{\rm WD}^{\rm PIN}$ ($M_\odot$)	$M_{\rm WD}^{\rm XIS}$ ($M_\odot$)
FO Aqr	$0.51^{+0.10}_{-0.08}$	$0.51^{+0.11}_{-0.07}$	EX Hya	$0.45^{+0.08}_{-0.06}$	$0.42^{+0.02}_{-0.01}$
XY Ari	$0.77^{+0.63}_{-0.21}$	$0.92^{+0.15}_{-0.14}$	NY Lup	$1.20^{+0.10}_{-0.20}$	$1.31^{+0.02}_{-0.03}$
MU Cam	$0.95^{+0.45}_{-0.35}$	$0.80^{+0.24}_{-0.14}$	V2400 Oph	$0.81^{+0.12}_{-0.10}$	$0.71^{+0.14}_{-0.15}$
BG CMi	$1.30^{+0.10}_{-0.54}$	$0.87^{+0.41}_{-0.29}$	AO Psc	$0.53^{+0.07}_{-0.06}$	$0.62^{+0.06}_{-0.11}$
V709 Cas	$1.10^{+0.14}_{-0.24}$	$0.75^{+0.40}_{-0.12}$	V1223 Sgr	$0.83^{+0.57}_{-0.17}$	$0.78^{+0.07}_{-0.09}$
TV Col	$0.87^{+0.53}_{-0.18}$	$0.91^{+0.09}_{-0.23}$	RX J2133	$1.10^{+0.15}_{-0.19}$	$0.80^{+0.16}_{-0.28}$
TX Col	$0.45^{+0.40}_{-0.17}$	$0.26^{+0.28}_{-0.03}$	IGR J17303	$1.20^{+0.15}_{-0.22}$	$1.05^{+0.18}_{-0.23}$
YY Dra	$1.40^{+0.00}_{-0.72}$	$0.68^{+0.27}_{-0.24}$	IGR J17195	$1.20^{+0.20}_{-0.40}$	$0.73^{+0.29}_{-0.07}$
PQ Gem	$1.10^{+0.16}_{-0.26}$	$0.75^{+0.65}_{-0.14}$			

In the previous subsection, we utilized these at the same time by fitting the XIS and HXD/PIN spectra simultaneously.

To check robustness of the wide-band fit, we also fitted the same XIS and HXD/PIN spectra separately. The XIS spectra were fitted with the same model of Eq. (5.2) as used in the wide-band fitting. Since the photo absorption process affects the shape of a spectrum only below ~10 keV, we applied the IP model (`IP_PSR`) directly to the HXD/PIN spectrum without photo absorption. The Fe abundance was fixed at unity (solar abundance) because the HXD/PIN spectra, containing no atomic emission lines, cannot constrain it. The derived WD masses are listed in Table 5.3, and plotted in Fig. 5.9. They are also compared with the mass estimations from the wide-band fittings in Fig. 5.10. The large errors are because of the low counting statistics in the HXD/PIN spectra, and the model degeneracy in the XIS band. Although, in Fig. 5.9, data points of some IPs (PQ Gem, V709 Cas, RX J2133, IGR J17195, BG CMi, and YY Dra) appear to systematically deviate from the dashed line ($y = x$ line), their two WD masses are still consistent each other if their large errors are considered. The remaining targets line up on the correlation line.

Thus, we have derived two important inferences. One is that the XIS and HXD/PIN spectra, which refer to the Fe-line ratio and the continuum shape, respectively, give consistent values of $M_{\rm WD}$. The other is that the combined wide-band fits give considerably smaller errors than the XIS-only or the PIN-only fits. These provide confidence in our analysis.

5.3 Systematic Uncertainties Involved in Our Analysis

In this section, we discuss systematic uncertainties which could be involved in the present analysis. As a whole, these may cause systematic uncertainties, and affect our WD mass results (Table 5.2) to some extent. Nevertheless, we regard the present

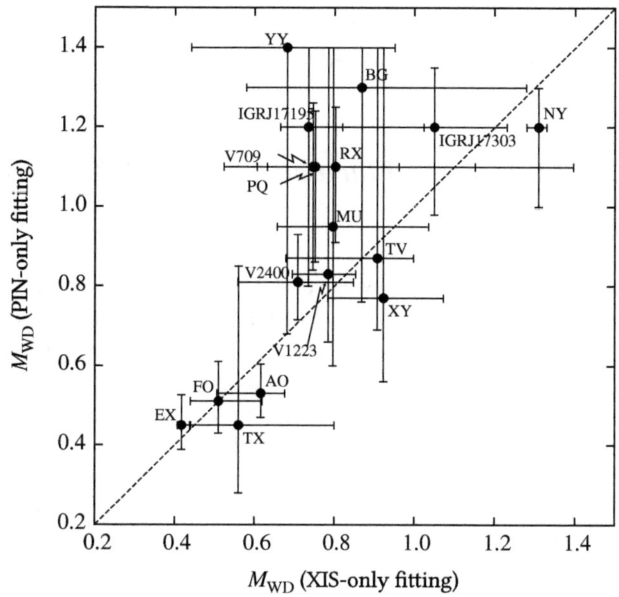

Fig. 5.9 Comparison of the WD masses estimated from the XIS-only and PIN-only fittings. Source names are abbreviated

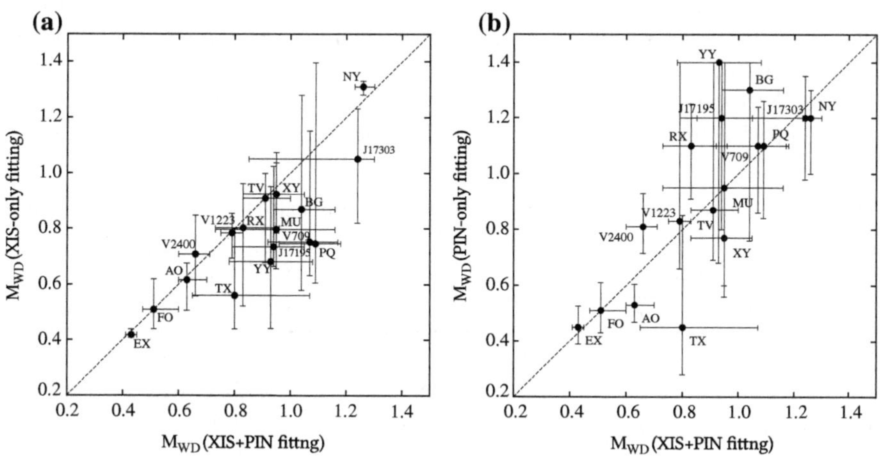

Fig. 5.10 Comparison of the WD masses estimated from the wide-band (XIS + PIN) fittings with those from **a** the XIS-only and **b** the PIN-only fittings

results as meaningful, because the estimated systematic effects are not significantly larger than the statistical precision associated with our determinations of $M_{\rm WD}$.

5.3.1 Effects of Systematic Fluctuations of the HXD/PIN Background

As noted in Sect. 4.1.2, background events of the HXD/PIN detector are estimated and subtracted using the background model which has a systematic uncertainty of ~1 % at 1σ (Fukazawa et al. 2009). This amounts to 0.005 counts $\rm s^{-1}$ in the HXD/PIN energy band, and corresponds to 1–10 % of the observed signal count rates (Table 5.1). Then, how does this uncertainty translate to that in the mass estimate?

One simple way to examine the above effect is to perform the same spectral fits using backgrounds of which the intensity is artificially increased (or decreased) by, for example, 1 %. Table 5.4 lists WD masses derived in this way. In many sources, the estimated WD masses are insensitive to the HXD/PIN NXB rescaling, resulting in similar WD masses (within 0.01–0.03 $M_\odot$), which is comparable to or smaller than the statistical fitting errors. Although TX Col shows a larger variation when changing the NXB normalization (0.70–0.91 $M_\odot$ with the HXD/PIN NXB rescaled by 0.99–1.01), the difference is still within the statistical fitting errors of this target. Therefore, we conclude that the systematic uncertainty of the HXD/PIN NXB is negligible in the present analysis. Note also that this *a posteiori* knowledge justifies validity of the target selection criteria, i.e., the hard X-ray flux above ~1 mCrab, explained in Sect. 5.1.

In the XIS spectrum, systematic uncertainties of the background subtraction is too small (~1 % in most energy bins) to affect the celestial signals.

Table 5.4 WD masses derived when the PIN background is rescaled by ±1 % which corresponds to its 1σ systematic uncertainties (Fukazawa et al. 2009)

System	NXB Rescaling			System	NXB Rescaling		
	0.99	1.00	1.01		0.99	1.00	1.01
FO Aqr	$0.51^{+0.09}_{-0.05}$	$0.51^{+0.09}_{-0.04}$	$0.51^{+0.09}_{-0.05}$	EX Hya	$0.43^{+0.02}_{-0.01}$	$0.43^{+0.02}_{-0.02}$	$0.42^{+0.02}_{-0.01}$
XY Ari	$0.96^{+0.08}_{-0.10}$	$0.95^{+0.10}_{-0.12}$	$0.92^{+0.12}_{-0.12}$	NY Lup	$1.26^{+0.05}_{-0.02}$	$1.26^{+0.04}_{-0.03}$	$1.26^{+0.04}_{-0.03}$
MU Cam	$0.95^{+0.16}_{-0.15}$	$0.95^{+0.21}_{-0.22}$	$0.95^{+0.18}_{-0.17}$	V2400 Oph	$0.67^{+0.06}_{-0.07}$	$0.66^{+0.05}_{-0.06}$	$0.64^{+0.07}_{-0.06}$
BG CMi	$1.04^{+0.12}_{-0.10}$	$1.04^{+0.12}_{-0.10}$	$1.04^{+0.13}_{-0.12}$	AO Psc	$0.68^{+0.05}_{-0.06}$	$0.63^{+0.07}_{-0.03}$	$0.63^{+0.05}_{-0.05}$
V709 Cas	$1.11^{+0.07}_{-0.19}$	$1.07^{+0.10}_{-0.15}$	$1.07^{+0.11}_{-0.17}$	V1223 Sgr	$0.79^{+0.04}_{-0.04}$	$0.79^{+0.04}_{-0.04}$	$0.79^{+0.04}_{-0.05}$
TV Col	$0.91^{+0.10}_{-0.08}$	$0.91^{+0.09}_{-0.08}$	$0.90^{+0.09}_{-0.09}$	RX J2133	$0.84^{+0.13}_{-0.10}$	$0.83^{+0.13}_{-0.10}$	$0.83^{+0.12}_{-0.11}$
TX Col	$0.91^{+0.37}_{-0.19}$	$0.80^{+0.27}_{-0.15}$	$0.70^{+0.21}_{-0.13}$	IGR J17303	$1.25^{+0.09}_{-0.37}$	$1.24^{+0.06}_{-0.39}$	$1.25^{+0.03}_{-0.35}$
YY Dra	$0.99^{+0.23}_{-0.20}$	$0.93^{+0.15}_{-0.15}$	$0.92^{+0.15}_{-0.21}$	IGR J17195	$0.94^{+0.16}_{-0.15}$	$0.94^{+0.11}_{-0.15}$	$0.93^{+0.09}_{-0.15}$
PQ Gem	$1.09^{+0.08}_{-0.12}$	$1.09^{+0.09}_{-0.13}$	$1.09^{+0.09}_{-0.14}$				

Results for a rescaling factor 1.00 are the same as listed in Table 5.2

5.3.2 Dynamics of Accreting Gas and System Geometries

As detailed in Suleimanov et al. (2005) and Brunschweiger et al. (2009), the assumption that the accreting gas falls freely from an infinite distance is not necessarily warranted, since the gas may well form an accretion disk and radiate away part of its energy release there. If such a disk is truncated, under plausible parameters, at a radius of $\sim 3 \times 10^9$ cm in an IP with a WD of 0.9 $M_\odot$, the shock temperature, and hence $M_{\rm WD}$, would be underestimated by $\sim$20 %.

Similarly, a fast spin of a WD would also cause the present method to underestimate $M_{\rm WD}$, because of the centrifugal force. If corrected for this effect, the WD masses in Table 5.2 of two rapid rotators in our sample, V709 Cas (with a spin period of 312.78 s) and YY Dra ($\sim$530 s), could increase by a few tens of percent depending on their accretion rates.

In addition to this, a fast spinning WD might obey a different (equatorially elongated) mass-radius relation (e.g., Nauenberg 1972; Geroyannis and Papasotiriou 1997). However, precise numerical treatment of this effect exceeds the scope of our present study, and we leave it untouched.

According to Canalle et al. (2005), the inclusion of the dipole curvature in the PSR calculation reduces the calculated shock temperature by $\sim$10 % for $M_{\rm WD} = 1.0\ M_\odot$ under typical accreting colatitudes, and <2–3 % for $M_{\rm WD} \lesssim 0.8\ M_\odot$. Since we cannot observationally constrain the magnetic field structure near the WD magnetic poles. Therefore, we neglected the curvature of the WD magnetic field when calculating the PSR structure (Chap. 3).

5.3.3 Accretion Rate Dependence

In the spectral fitting, we freely varied the normalization (i.e., accretion rate) of the PSR spectral model, which was calculated assuming a fixed specific accretion rate of 1.0 g cm^{-2} s^{-1} (Sect. 3.3). Strictly speaking, this means a self-inconsistency, because the calculated shock height and the shock temperature should increase and decrease (due to the reduced potential depth), respectively, as the accretion rate decreases; this, in turn, is because the bremsstrahlung cooling rate is proportional to the plasma density squared (e.g., Aizu 1973).

The total mass accretion rate can be computed from the derived WD mass and the observed fluxes (Table 5.2) if the distance to each object is specified. Further using the fractional area f of a PSR footprint to the WD surface, we can convert the value into the specific mass accretion rate a measured at the shock height (note that we assumed a cylindrical PSR structure). Table 5.5 presents results of mass accretion rates calculated in this way. Distances are taken from references listed in the table. In the calculation, we fixed f at 0.001 as the same value used in the present model calculation (Sect. 3.3). The resulting mass accretion rates range over 0.2–23.2 g cm^{-2} s^{-1}, which are very typical values for IPs (e.g., Patterson 1984, 1994). Thus, these values do not precisely equal 1 assumed above.

Table 5.5 Mass accretion rates calculated based on the WD masses derived from the present spectral analysis (Table 5.2)

System	Assumptions		Resultant values	
	M_{WD}	d [a]	$\dot{M}$ [b]	a [c]
	$M_\odot$	pc	$10^{-11} M_\odot$/yr	g cm^{-2} s^{-1}
FO Aqr	0.51	400[1]	118.5	6.6
XY Ari	0.95	270[2]	6.1	0.9
BG CMi	1.04	700[3]	36.8	7.0
TV Col	0.91	386[4]	29.6	4.0
TX Col	0.80	550[5]	15.8	1.6
YY Dra	0.93	155[6]	2.2	0.3
EX Hya	0.43	64.5[7]	4.8	0.2
NY Lup	1.26	690[8]	45.4	21.1
AO Psc	0.63	420[9]	59.7	4.3
V1223 Sgr	0.79	527[10]	229.2	23.2

As described in Sect. 3.3, the footprint of an accretion column is assumed to be 0.1% (i.e. $f = 0.001$) of the WD surface area. Distances to each system are retrieved from the publications listed below
[a] References: [1] McHardy et al. (1987), [2] Littlefair et al. (2001), [3] Berriman (1987), [4] McArthur et al. (2001), [5] Buckley and Tuohy (1989), [6] Mateo et al. (1991), [7] Beuermann et al. (2003), [8] de Martino et al. (2006), [9] Hellier et al. (1991), [10] Beuermann et al. (2004)
[b] Total mass accretion rate
[c] Specific mass accretion rate measured at the shock height

To examine the accretion-rate dependence of our numerical result, and hence that of the derived WD masses, we performed calculations with several different accretion rates as shown in Fig. 5.11. As noted above, a lower value of a gives a lower shock temperature at any M_{WD}. At higher M_{WD}, differences of the resulting shock temperature becomes more conspicuous. At $M_{WD} = 0.8\ M_\odot$, we see a difference by a factor of 1.13 between $a = 0.1$ and 1.0 g cm^{-2} s^{-1}, and a factor of 1.02 between $a = 1.0$ and 10 g cm^{-2} s^{-1}. Correspondingly, at $M_{WD} = 1.2\ M_\odot$, the difference increases to factors of 1.33 ($a = 0.1$ to $a = 1.0$) and 1.10 ($a = 1.0$ to $a = 10$). In most systems we analyzed, however, the WD masses are below 1.2 $M_\odot$, so we consider that this uncertainty is considered to affect the WD mass estimates only slightly, much less than ~30 %. Especially in the mass regime below 1 $M_\odot$, the a dependence is so small that the estimated shock temperatures and WD masses are considerably secure.

5.3.4 X-ray Reflection on the WD Surface

The spectra of the present IP sample clearly exhibit the fluorescent Fe line at ~6.4 keV. This implies that some of the incident X-rays emitted from the PSR are reflected by cold gas near the WD surface or pre-shock region (see Ezuka and Ishida 1999, and references therein). However, in the present spectral fitting (Sect. 5.2.2) we did not take into account a reflection component from the WD surface explicitly, because

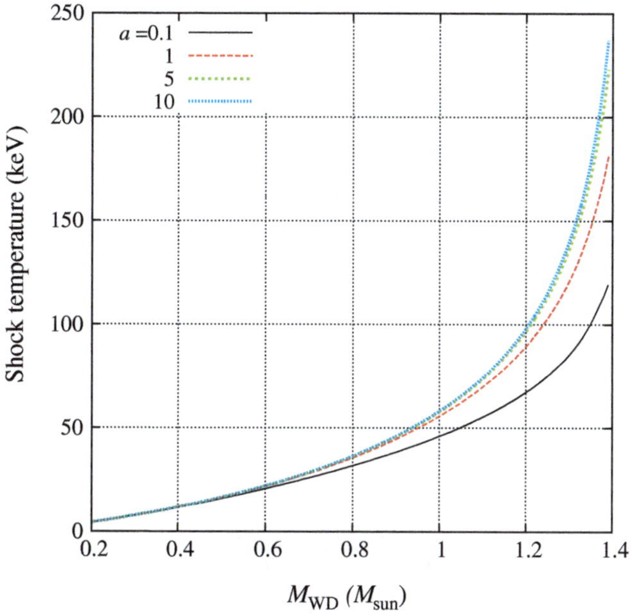

Fig. 5.11 Results of the numerical calculation of the PSR model when changing the accretion rate. Shock temperatures are plotted as *black solid*, *red long dashed*, *green short dashed*, and *blue dotted lines* for accretion rates of 0.1, 1, 5, and 10 g cm^{-2} s^{-1}, respectively

the computation power needed to precisely calculate the reflection is unaffordable at this moment. Qualitatively, its effect may be represented to some extent by the strongly absorbed component in the partial-covering model, which has a similar spectral shape to a reflection component, both contributing mainly in the HXD/PIN energy band.

Cropper et al. (1998) revealed that an inclusion of a reflection component changes the obtained WD masses only slightly from those of the partial covering modeling. By analyzing a wide-band spectrum of V1223 Sgr with a reflection component, Revnivtsev et al. (2004) derived a WD mass of $0.71 \pm 0.03\ M_{\odot}$. Our result, $0.75 \pm 0.05\ M_{\odot}$ well agrees with their value, and thus an effect of the reflection component can be considered relatively small.

5.4 Discussion

5.4.1 WD Mass Spectrum and Its Average

Figure 5.12 shows the distribution of the WD mass of our sample, compared with those of CVs determined kinematically, mainly based on optical/infrared spectroscopy (Ritter and Kolb 2003). The mean mass is obtained as $0.88 \pm 0.24\ M_{\odot}$

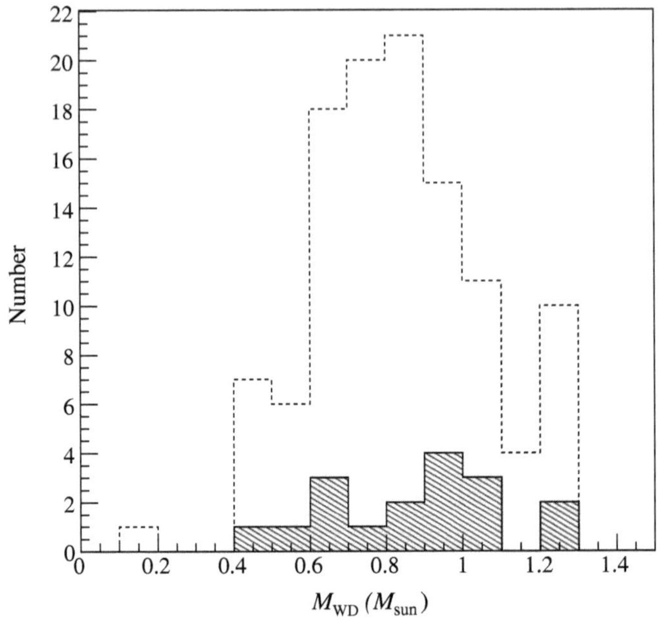

Fig. 5.12 The WD mass spectrum of our sample IPs (*shaded histogram*), compared with that of 104 CVs (*histogram with dashed line*; taken from Ritter and Kolb 2003), derived from optical/infrared spectroscopies

from the present study and $0.82 \pm 0.23\ M_{\odot}$ from the optical sample (errors are 1σ standard deviations), with a Kolmogorov-Smirnov probability of 0.39. Therefore it is inferred for this limited sample that the mass spectra of CVs and IPs are statistically not largely different.

The mean WD mass of our IP sample, $0.88 \pm 0.24\ M_{\odot}$, can be compared with the value of $\sim 0.5\ M_{\odot}$ inferred by Krivonos et al. (2007) based on the analysis of Galactic Ridge X-ray Emission (GRXE) in the hard X-ray band with *INTEGRAL*/SPI. In Chap. 6, we analyze our own data of the GRXE, and fit them using the IP emission model, and discuss derived WD masses.

5.4.2 Comparison with Measurements of WD Masses

5.4.2.1 Results Using X-ray Data

Table 5.6 compares the present *Suzaku* XIS + HXD results with the WD masses reported by the previous studies in the X-ray band. In Fig. 5.13, we plot correlations between our WD masses and those from the *RXTE*/PCA + HEXTE data by Suleimanov et al. (2005) and the *Swift*/BAT data by Brunschweiger et al. (2009); the spectral models and the energy band used in these three studies are very similar to one another.

Table 5.6 Estimated WD masses compared with previous reports in the X-ray band. The mass ranges associated are all 90 % confidence intervals

System	*Suzaku*[a] XIS + HXD $M_{\rm WD}$	*Swift*[b] BAT $M_{\rm WD}$	*RXTE*[c] PCA + HEXTE $M_{\rm WD}$	*RXTE*[d] PCA $M_{\rm WD}$	*Ginga*[e] LAC $M_{\rm WD}$	*ASCA*[f] SIS $M_{\rm WD}$
FO Aqr	$0.51^{+0.09}_{-0.04}$	0.61 ± 0.05	0.60 ± 0.05	$0.88^{+0.07}_{-0.09}$	$0.92^{+0.30}_{-0.34}$	$1.05(> 0.44)$
XY Ari	$0.95^{+0.10}_{-0.12}$	0.96 ± 0.12		$0.97^{+0.16}_{-0.17}$		
MU Cam	$0.95^{+0.21}_{-0.22}$	0.74 ± 0.13				
BG CMi	$1.04^{+0.12}_{-0.10}$	0.67 ± 0.19	0.85 ± 0.12	$1.15^{+0.06}_{-0.06}$	$1.09(> 0.94)$	
V709 Cas	$1.07^{+0.10}_{-0.15}$	0.96 ± 0.05	0.90 ± 0.10	$1.08^{+0.05}_{-0.17}$		
TV Col	$0.91^{+0.09}_{-0.08}$	0.78 ± 0.06	0.84 ± 0.06	$0.97^{+0.02}_{-0.05}$	$1.3(> 0.9)$	$0.51^{+0.41}_{-0.22}$
TX Col	$0.80^{+0.27}_{-0.15}$	0.67 ± 0.10	0.70 ± 0.30	$0.74^{+0.05}_{-0.05}$	$0.48^{+0.06}_{-0.09}$	$0.66^{+0.73}_{-0.42}$
YY Dra	$0.93^{+0.15}_{-0.15}$	0.50 ± 0.11	0.75 ± 0.05			
PQ Gem	$1.09^{+0.09}_{-0.13}$	0.65 ± 0.09	0.65 ± 0.20		$1.29(> 1.17)$	
EX Hya	$0.43^{+0.02}_{-0.02}$	0.66 ± 0.17	0.50 ± 0.05	$0.44^{+0.03}_{-0.03}$	$0.46^{+0.04}_{-0.04}$	$0.48^{+0.10}_{-0.06}$
NY Lup	$1.26^{+0.04}_{-0.03}$	1.09 ± 0.07				
V2400 Oph	$0.66^{+0.05}_{-0.06}$	0.81 ± 0.10	0.59 ± 0.05	$0.71^{+0.07}_{-0.03}$		$0.68^{+0.42}_{-0.24}$
AO Psc	$0.63^{+0.07}_{-0.03}$	0.55 ± 0.06	0.65 ± 0.05	$0.60^{+0.03}_{-0.03}$	$0.56^{+0.16}_{-0.20}$	$0.40^{+0.13}_{-0.10}$
V1223 Sgr	$0.79^{+0.04}_{-0.04}$	0.65 ± 0.04	0.95 ± 0.05	$1.07^{+0.08}_{-0.09}$		$1.28(> 0.84)$
RX J2133	$0.83^{+0.13}_{-0.10}$	0.91 ± 0.06				
IGR J17303	$1.24^{+0.06}_{-0.39}$	1.08 ± 0.07				

[a]This work,
[b]Brunschweiger et al. (2009),
[c]Suleimanov et al. (2005),
[d]Ramsay (2000),
[e]Cropper et al. (1999),
[f]Ezuka and Ishida (1999)

In Fig. 5.13a and b, our results derive higher WD masses, especially for massive ($\gtrsim 1\ M_\odot$) systems, although rough correlations are seen among the three measurements. For a few sources, for example V709 Cas, BG CMi, and PQ Gem, this discrepancy amounts to ~0.3–0.4 $M_\odot$. Reasons for these differences are unclear at present. Unfortunately, the WD masses of Suleimanov et al. (2005) and Brunschweiger et al. (2009) show the same level of disagreement as seen in these figures, even though the two studies use the same spectral model calculated in the former. Figure 5.14 represents this more clearly. The difference within the same model might indicate that the results suffer from rather large systematic uncertainties of different instruments (*RXTE*/PCA and *Swift*/BAT), which could exceed the quoted statistical fitting errors. We are aware that our analysis is not free, either from possible systematic uncertainties (Sect. 5.3) involved in our instruments (*Suzaku*/XIS + HXD) and the spectral model. Yet we think that our result may be less subject to various systematic

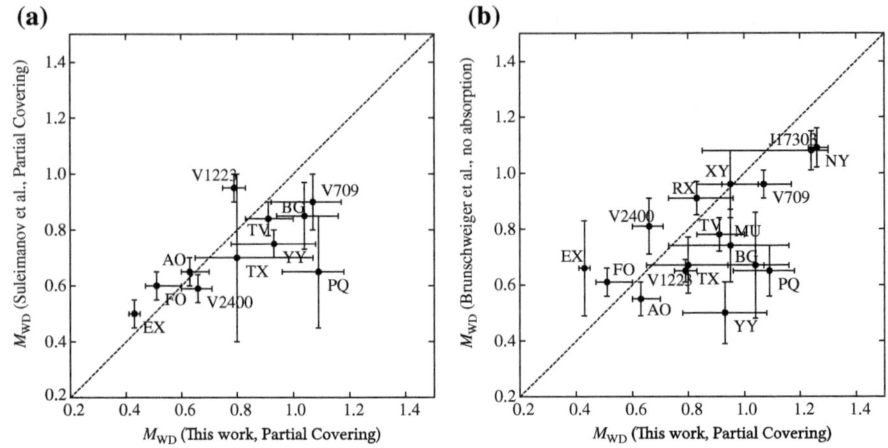

Fig. 5.13 Comparison of the WD masses estimated in the present study with those by **a** Suleimanov et al. (2005) and **b** Brunschweiger et al. (2009)

Fig. 5.14 Comparison of the WD masses reported by Suleimanov et al. (2005) and Brunschweiger et al. (2009), both using the same spectral model but different instruments in the fittings

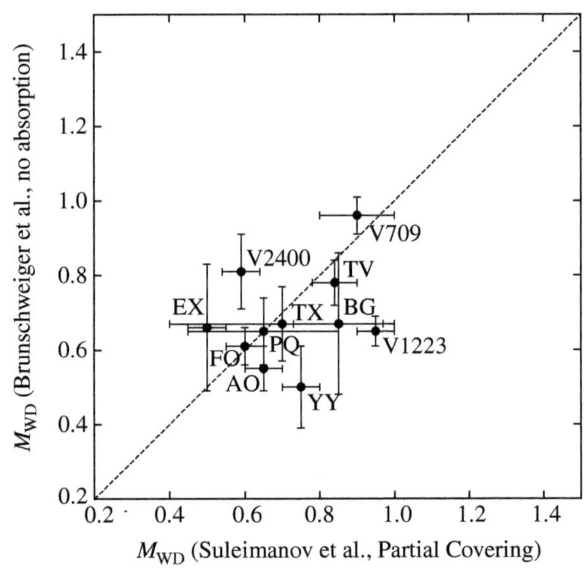

effects, because we jointly utilized the Fe-K line spectroscopy and the hard-band continuum shape. It also supports this idea that our three different fitting strategy, the wide-band (XIS + PIN), the XIS-only, and the PIN-only fittings gave the consistent WD masses (Fig. 5.9).

In Yuasa et al. (2010), we have already reported WD mass estimates based on the same *Suzaku* data but with slightly different spectral model. In that paper, a plasma cooling function of Sutherland and Dopita (1993) was used when solving

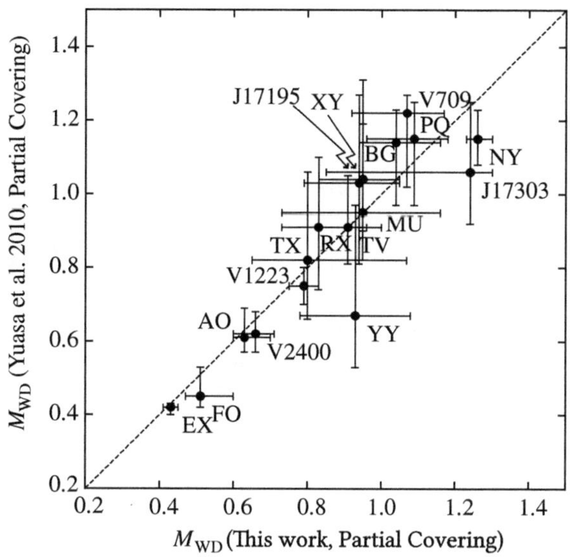

Fig. 5.15 Comparison of the WD masses estimated in the present study with those Yuasa et al. (2010)

PSR equations, and hence the Fe abundance had to be fixed at 1. In contrast, the present model utilizes an update cooling function of Schure et al. (2009) which also allows variable Fe abundance (Chap. 3). As shown in Fig. 5.15, the two results agree well within statistical fitting errors, and thus, it is clear that the WD masses estimated in the present method do not depend on the choice among the currently used cooling functions. Note, however, that the presently updated modeling has higher self-consistency and physical accuracy compared to that of Yuasa et al. (2010).

From X-ray light curve analysis in the eclipsing IP XY Ari, Hellier (1997) estimated M_{WD} to be 0.91–1.17 $M_{\odot}$. Our result, $M_{WD} = 0.95^{+0.10}_{-0.12}\ M_{\odot}$, nicely agrees with this value, and this agreement provides a convincing calibration to the present methodology (see also Brunschweiger et al. 2009 for a similar argument).

5.4.2.2 Results Using Other Methods

It is also meaningful to compare the derived WD masses with those estimated from binary kinematics based mainly on optical/IR spectroscopy. Using references in the CV catalog by Ritter and Kolb (2003), we collected previously reported M_{WD} of IPs, and then plotted them as shown in Fig. 5.16. Note that most of the reports employed an assumption of a Roche-lobe filling Zero Age Main Sequence star to estimate the radius and mass of the secondary (non-degenerate) stars, and this may introduce a cause of systematic uncertainties in their WD mass estimations.

All masses agree (with rather large errors) except for EX Hya. For this object, WD masses determined spectroscopically in the optical/infrared wavelength have been somewhat controversial, mainly reflecting different radial velocity estimations. Based on Na I and Ca II line profiles, Beuermann and Reinsch (2008) estimated

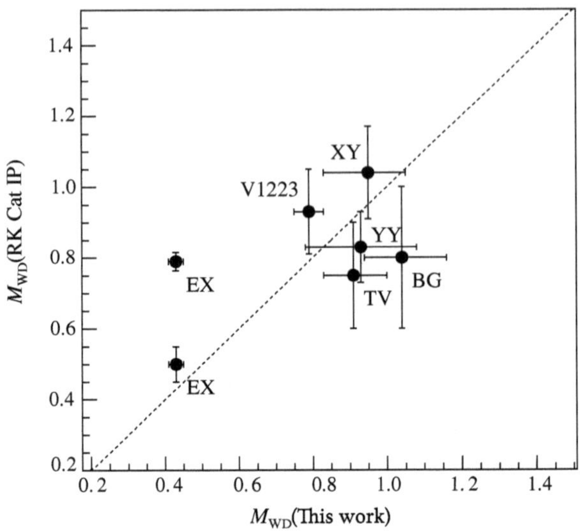

Fig. 5.16 Comparison of the WD masses estimated in the present study with those in the optical and infra-red wavelengths. References are Hellier (1997) for XY Ari, Penning (1985) for BG CMi, Hellier (1993) for TV Col, Haswell et al. (1997) for YY Dra, and Beuermann et al. (2004) for V1223 Sgr. For EX Hya, two different values are presented; $0.790 \pm 0.026\ M_{\odot}$ and $0.50 \pm 0.05\ M_{\odot}$ by Beuermann and Reinsch (2008) and Mhlahlo et al. (2007), respectively (see text). Quoted errors all refer to 90% confidence levels

the amplitude of radial velocity of the secondary (non-degenerate) star K_2 to be 432.4 ± 4.8 km s^{-1} by assuming that of the primary WD $K_1 = 59 \pm 3$ km s^{-1} (Belle et al. 2003; Hoogerwerf et al. 2004). From these values, they obtained $M_{\rm WD} = 0.790 \pm 0.026\ M_{\odot}$. On the contrary, Mhlahlo et al. (2007) derived $K_1 = 74 \pm 2$ km s^{-1} from Hα, Hβ, and Hγ lines, and used K_2 of 360 ± 35 km s^{-1} reported by Vande Putte et al. (2003), leading to the WD mass $M_{\rm WD} = 0.50 \pm 0.05 M_{\odot}$. This disagrees with the former value by more than four times the error range, but is closer to our measurement.

The HXD/PIN spectrum of EX Hya clearly shows a cutoff in the 10–20 keV band (Fig. A.1), and because of this cutoff together with the Fe line structure, a lower WD mass is favored by the fitting. The cutoff cannot be explained by a WD of 0.7 $M_{\rm WD}$, since our model would then require a shock temperature of 30 keV and a similar cutoff-energy; multiple absorption or reflection would make the cutoff even higher than this value.

5.4.3 Comparison of Fe Abundances

Fujimoto and Ishida (1997), Ezuka and Ishida (1999), and Evans et al. (2004) estimated metal abundance of IPs by analyzing the resolved Fe line complex (6–7 keV) and X-ray continua with the multi-temperature PSR model. Fujimoto and Ishida

Table 5.7 The derived Fe abundances of accreting plasmas of IPs compared with ones reported by Ezuka and Ishida (1999). Associated are 90 % confidence intervals

System	This work $Z_{\rm Fe}(Z_\odot)$	Ezuka and Ishida $Z_{\rm Fe}(Z_\odot)$
FO Aqr	0.10(0.00–0.12)	0.48(0.33–0.64)
XY Ari	0.36(0.33–0.39)	0.10(0.01–0.23)
TV Col	0.41(0.38–0.45)	0.31(0.22–0.41)
TX Col	0.41(0.32–0.50)	0.52(0.30–0.78)
PQ Gem	0.17(0.14–0.22)	0.10(0.05–0.21)
EX Hya	0.55(0.52–0.59)	0.72(0.60–0.80)
V2400 Oph	0.25(0.21–0.28)	0.39(0.30–0.48)
AO Psc	0.46(0.42–0.53)	0.41(0.30–0.55)
V1223 Sgr	0.27(0.25–0.27)	0.68(0. 54–0.83)

(1997) obtained the Fe abundance of EX Hya as $0.6 \pm 0.2\ Z_\odot$, which was later revised by Ezuka and Ishida (1999) to $0.72^{+0.08}_{-0.12}\ Z_\odot$ together with measurements of Fe abundances of 12 other IPs. Evans et al. (2004) gave the metal abundance of $0.121^{+0.028}_{-0.023}\ Z_\odot$ for FO Aqr (whilst they did not take their result as evidence of genuine sub-solar abundance).

Since our IP sample is well covered by that of Ezuka and Ishida (1999), we compare our result with theirs in Table 5.7 and Fig. 5.17. Although the two results consistently indicate sub-solar abundance, they exhibit no particular correlation. Considering that the *Suzaku* data have a much wider spectral coverage, higher counting statistics, and a somewhat better energy resolution than those from *ASCA* which Ezuka and Ishida (1999) used, our results may be seen as more updated. Note also that Ezuka and Ishida (1999) did not take into account partial covering or the reflection component.

When our 17 objects are divided into five objects with $M_{\rm WD} > 1.0\ M_\odot$ and the other twelve with $M_{\rm WD} \leq 1.0\ M_\odot$, the former and latter sub-samples have averaged Fe abundances of $0.23 \pm 0.10\ Z_\odot$ and $0.38 \pm 0.13\ Z_\odot$, respectively. The lack of significant difference ensures that our abundance determinations are not strongly affected by the continuum shape or the WD mass.

Previous studies of summed, or unresolved (due to lower energy resolution), Fe lines also reported sub-solar Fe abundances in IPs. Using *Ginga*, Beardmore et al. (2000) reported an under abundance of metals by a factor of ~2 relative to solar for V1223 Sgr. This roughly agrees with our result, 0.30–0.35 $Z_\odot$. de Martino et al. (2004) gave $0.33^{+0.07}_{-0.09}\ Z_\odot$ for PQ Gem (RE0751 + 14), which is also comparable to ours (0.19–0.32 $Z_\odot$).

In magnetic CVs with special accreting and viewing geometries, the Fe emission line (6.7 keV) could be artificially enhanced or suppressed by "resonance trapping" of the Fe-K line photons (Terada et al. 2001), because the optical depth of this process is higher than 1 (e.g., 25 by Done et al. 1995) in a typical PSR. This could affect our abundance measurements, and in turn, those of $M_{\rm WD}$. However, this effect is only of importance when the column is observed either pole-on (Fe-K line enhanced) or

Fig. 5.17 Comparison of the Fe abundances estimated in the present wide-band study with those from an *ASCA* study of the Fe emission lines (Ezuka and Ishida 1999)

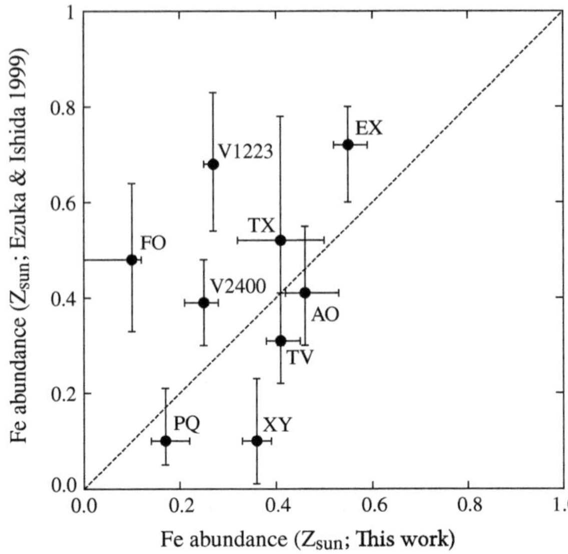

side-on (suppressed) through a large fraction of the spin phase (Terada et al. 2001). Because our 17 IPs are more or less pulsating at their spin periods, they are not likely to be either pole-on or side-on, and the enhancement or suppression, if any, would be averaged out by the WD spin (changing viewing angle). We therefore consider that this effect is negligible for the present study, and that the derived sub-solar abundances reflect a true property of the accreting gases.

Finally, we may compare our abundance values with those derived from nearby isolated stars using optical surveys. For example, Nordström et al. (2004) reported a mean metal abundance of 0.72 $Z_\odot$ ($\log_{10}$[Metal/H] $= -0.14$) for F- and G-type stars. The value is as twice as high compared to the Fe abundance we derived for the IP sample. Although we generally assume that low-mass late-type stars contained in IPs do not largely differ from isolated ones, the suggested difference may imply some systematic effect working in our analysis method or the binary evolution path.

5.5 Summary of the WD Mass Estimation

By numerically solving the hydrostatic equations for post-shock regions on top of magnetic CVs (e.g., Suleimanov et al. 2005), we constructed the X-ray spectral model with improvements in terms of (1) a plasma cooling function with variable heavy metal abundance, and (2) the optically-thin thermal emission code including the Fe emission lines in place of simple Bremsstrahlung (Chap. 3). We fitted spectra of 17 IPs observed with *Suzaku*, and successfully reproduced the Fe line structures in the 6–7 keV band and at the same time the thermal cutoff detected in the hard X-ray band (Sect. 5.2.2). From the best-fit models, we estimated the WD masses and the Fe abundances. The average WD mass for the present sample were $0.88 \pm 0.23\ M_{\odot}$. All sources exhibited sub-solar Fe abundance, with the average being $0.33 \pm 0.14\ Z_{\odot}$. The WD mass of the newly found IP, IGR J17195−4100, was estimated to be $0.94^{+0.11}_{-0.15}\ M_{\odot}$.

We examined various systematic uncertainties which might affect the obtained WD masses (Sect. 5.3), and conclude that none of them can change the result significantly when compared to statistical fitting errors associated with the mass estimations. The derived WD masses are compared with previous reports (Sect. 5.4.2), and a rough correlation was confirmed, although results on a few systems differ significantly ($\sim$0.3–0.4 $M_{\odot}$). Considering the improvements in the model construction and the higher spectral quality, the present result is thought to have better accuracy than the previous reports. This statement is supported in particular by the combined use of the Fe-K emission line ratios (with the XIS) and the hard continuum shapes (with the HXD), enabled for the first time with *Suzaku*.

References

Aizu, K. 1973, Progress of Theoretical Physics, 49, 1184

Beardmore, A. P., Osborne, J. P., & Hellier, C. 2000, MNRAS, 315, 307

Belle, K. E., Howell, S. B., Sion, E. M., Long, K. S., & Szkody, P. 2003, Apj, 587, 373

Berriman, G. 1987, A&AS, 68, 41

Beuermann, K., Harrison, T. E., McArthur, B. E., Benedict, G. F., & Gänsicke, B. T. 2003, A&A, 412, 821

Beuermann, K., Harrison, T. E., McArthur, B. E., Benedict, G. F., & Gänsicke, B. T. 2004, A&A, 419, 291

Beuermann, K. & Reinsch, K. 2008, A&A, 480, 199

Boldt, E. 1987, Phys. Rep., 146, 215

Brunschweiger, J., Greiner, J., Ajello, M., & Osborne, J. 2009, A&A, 496, 121

Buckley, D. A. H. & Tuohy, I. R. 1989, Apj, 344, 376

Canalle, J. B. G., Saxton, C. J., Wu, K., Cropper, M., & Ramsay, G. 2005, A&A, 440, 185

Cropper, M., Ramsay, G., & Wu, K. 1998, MNRAS, 293, 222

Cropper, M., Wu, K., Ramsay, G., & Kocabiyik, A. 1999, MNRAS, 306, 684

de Martino, D., Bonnet-Bidaud, J., Mouchet, M., et al. 2006, A&A, 449, 1151

de Martino, D., Matt, G., Belloni, T., et al. 2004, Nuclear Physics B Proceedings Supplements, 132, 693

Done, C. & Magdziarz, P. 1998, MNRAS, 298, 737

Done, C., Osborne, J. P., & Beardmore, A. P. 1995, MNRAS, 276, 483

Evans, P. A., Hellier, C., Ramsay, G., & Cropper, M. 2004, MNRAS, 349, 715

Ezuka, H. & Ishida, M. 1999, ApJs, 120, 277

Fujimoto, R. & Ishida, M. 1997, Apj, 474, 774

Fukazawa, Y., Mizuno, T., Watanabe, S., et al. 2009, PASJ, 61, 17

Geroyannis, V. S. & Papasotiriou, P. J. 1997, Ap&SS, 249, 279

Haberl, F., Motch, C., & Zickgraf, F. 2002, A&A, 387, 201

Haswell, C. A., Patterson, J., Thorstensen, J. R., Hellier, C., & Skillman, D. R. 1997, ApJ, 476, 847

Hellier, C. 1993, MNRAS, 264, 132

Hellier, C. 1997, MNRAS, 291, 71

Hellier, C., Cropper, M., & Mason, K. O. 1991, MNRAS, 248, 233

Hellier, C. & Mukai, K. 2004, MNRAS, 352, 1037

Hoogerwerf, R., Brickhouse, N. S., & Mauche, C. W. 2004, ApJ, 610, 411
Ishisaki, Y., Maeda, Y., Fujimoto, R., et al. 2007, PASJ, 59, 113
Kokubun, M., Makishima, K., Takahashi, T., et al. 2007, PASJ, 59, 53
Krivonos, R., Revnivtsev, M., Churazov, E., et al. 2007, A&A, 463, 957
Littlefair, S. P., Dhillon, V. S., & Marsh, T. R. 2001, MNRAS, 327, 669
Mateo, M., Szkody, P., & Garnavich, P. 1991, ApJ, 370, 370
McArthur, B. E., Benedict, G. F., Lee, J., et al. 2001, ApJ, 560, 907
McHardy, I. M., Pye, J. P., Fairall, A. P., & Menzies, J. W. 1987, MNRAS, 225, 355
Mhlahlo, N., Buckley, D. A. H., Dhillon, V. S., et al. 2007, MNRAS, 378, 211
Nauenberg, M. 1972, ApJ, 175, 417
Nordström, B., Mayor, M., Andersen, J., et al. 2004, A&A, 418, 989
Patterson, J. 1984, ApJs, 54, 443
Patterson, J. 1994, PASP, 106, 209
Penning, W. R. 1985, ApJ, 289, 300
Ramsay, G. 2000, MNRAS, 314, 403
Revnivtsev, M., Lutovinov, A., Suleimanov, V., Sunyaev, R., & Zheleznyakov, V. 2004, A&A, 426, 253
Ritter, H. & Kolb, U. 2003, A&A, 404, 301
Schure, K. M., Kosenko, D., Kaastra, J. S., Keppens, R., & Vink, J. 2009, A&A, 508, 751
Serlemitsos, P. J., Soong, Y., Chan, K.-W., et al. 2007, PASJ, 59, 9
Suleimanov, V., Revnivtsev, M., & Ritter, H. 2005, A&A, 435, 191
Sutherland, R. S. & Dopita, M. A. 1993, ApJS, 88, 253
Terada, Y., Ishida, M., Makishima, K., et al. 2001, MNRAS, 328, 112
Vande Putte, D., Smith, R. C., Hawkins, N. A., & Martin, J. S. 2003, MNRAS, 342, 151
Yuasa, T., Nakazawa, K., Makishima, K., et al. 2010, A&A, 520, A25

Chapter 6
Decomposing the Galactic Ridge X-ray Emission

In this chapter, we perform detailed spectroscopy of the Galactic Ridge X-ray Emission (GRXE). A wide-band high quality *Suzaku* spectrum of the GRXE is fitted with a model which consists of several model components each of which corresponds to a class of low-luminosity X-ray point source. Based on fitting results, the origins of the GRXE are discussed.

6.1 Observation and Preparation of Data Sets

6.1.1 Data Selection

Since the launch, *Suzaku* has been actively observing the Galactic center and the Galactic bulge region ($|l| \lesssim 5°$, $|b| \lesssim 3°$) to study the extreme environment around the supermassive black hole, supernovae, and molecular clouds reflecting X-rays; this has become one of major key scientific topics of *Suzaku*. Figure 6.1a shows a mosaic X-ray image of the Galactic center region constructed by collecting all these *Suzaku* data. The data used to create this mosaic are taken from 92 mapping observations performed during 2005–2010 of which the individual log are tabulated for reference in Appendix B. The data of these observations contain plenty of signals from the GRXE because of its deep exposure (see below), and therefore very suitable for analyzing spectra of this emission. However, data used in the analysis should be selected with caution because, in some observations, bright X-ray point sources fall inside the instrument fields of view, and contaminate the GRXE signal making the data useless. In fact, typical fluxes from bright X-ray sources in this region exceed that of the GRXE by 1–2 orders of magnitude in the HXD energy band for example.

To avoid the bright source contamination, we carefully selected observations that do not contain any bright point source. Point sources listed in the *INTEGRAL* General Reference Catalog (Ebisawa et al. 2003) were used for referring to positions and typical intensities of such sources as marked with circles in Fig. 6.1. We find that

T. Yuasa, Suzaku *Studies of White Dwarf Stars and the Galactic X-ray Background Emission*, Springer Theses, DOI: 10.1007/978-4-431-54219-3_6, © Springer Japan 2013

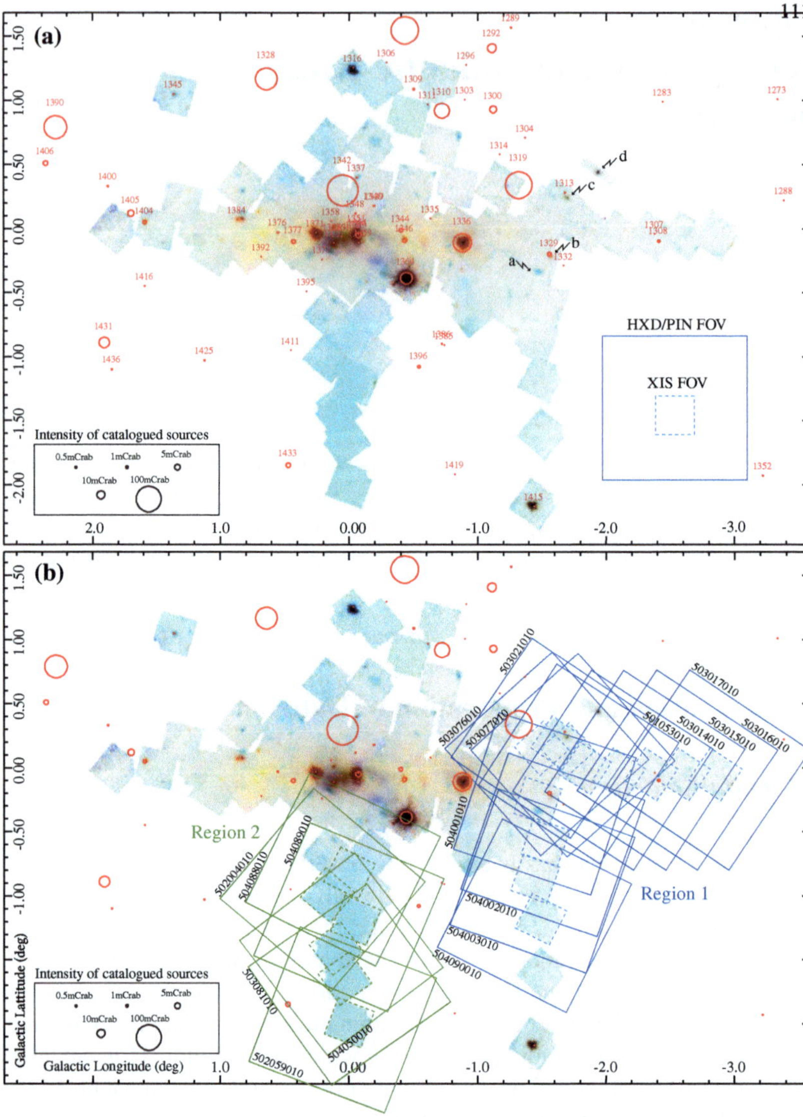

Fig. 6.1 **a** A composite three-color X-ray image of the Galactic center and the Galactic bulge region created by adding data of 92 mapping observations performed during 2005–2010. It was constructed from the XIS0, 1, and 3 data in the 0.5–2, 2–5, and 5–8 keV bands, which are represented by *blue*, *green*, and *red*, respectively. The image is corrected for the vignetting effect of the XRT and exposure difference, and the NXB was subtracted. Each *rectangular* image corresponds to a single observation. *Circles* show the position and intensity of known X-ray sources listed in the *INTEGRAL* General Reference Catalog (Ebisawa et al. 2003), accompanied by their IDs in the catalog (Table 6.2). *Labeled arrows* indicate contaminating point sources analyzed in Sect. 6.1.2. The fields of view of the XIS and HXD/PIN are presented for reference. **b** The same image as (**a**) but the detector fields of view of the selected observations are overlaid

the actual XIS signal strength of some sources differ from their cataloged intensities which are represented by red circles in Fig. 6.1; this is probably because they are transient (i.e., highly time variable). For example, the a red circle for the hard X-ray transient EXS 1737.9−2952 (source ID 1319; Grindlay et al. 1993) is rather large, but it is not detected in the XIS image. The source was only bright in the past, and very faint when *Suzaku* observed it. When a certain observation is contaminated by X-ray point sources of which the total flux is above 0.1 mCrab (equivalently $\sim 2 \times 10^{-12}$ erg cm^{-2} s^{-1} in the 2–8 keV) inside the HXD field of view, we discarded the data set. This criteria is set because a GRXE flux observed with the HXD is typically 1–2 mCrab in this sky region, and the limiting flux 0.1 mCrab corresponds to 5–10 % of the detected flux. Mostly acceptable ones were observations of (1) the Galactic plane in negative longitudes, and (2) off-plane regions with negative latitudes. Table 6.1 lists observations which passed the criteria, and are used in the present spectral analysis. In Fig. 6.1b, the fields of view of the utilized observations are overlaid on the same X-ray image as panel (a). To study spatial variations of spectral properties of the GRXE, we divided these observations into two groups, and hereafter refer to them as Region 1 (on-plane, $l < 0$ regions) and Region 2 (off-plane, $b < 0$ regions).

Compared to typical *Suzaku* observations of other types of sources, the total effective exposures of the selected regions are extremely deep, 592.1 ks (XIS) and 524.0 ks (HXD/PIN) for Region 1, and 418.8 ks (XIS) and 340.2 ks (HXD/PIN) for Regions 2. Owing to these exposures, the extracted spectra have high counting statistics, and cover the wide energy band over 2–50 keV as explained in Sect. 6.1.5. This energy coverage is crucial for clarifying the nature of the GRXE. As described below, this data set also achieves the highest quality, in terms of absence of contamination from bright point sources, that has ever been available from all the past and current hard X-ray missions, including *BeppoSAX*, *RXTE*, *INTEGRAL*, and *Swift*.

6.1.2 Contaminating Fluxes from Apparent Point Sources

Although we selected observations with very low point-source contamination, some sources are still seen in the XIS images, and therefore, their influence on our GRXE study should be assessed. In the following paragraphs, we evaluate their fluxes, and eventually conclude that they can be negligible in the present GRXE analysis in the hard X-ray band.

As indicated with black arrows in Fig. 6.1, we note four faint X-ray point sources in the fields of view of Region 1. Source a is a newly-found soft X-ray source, and the remaining three are previously known sources; b = AX J1742.6−3022 (Sakano et al. 2002), c = Suzaku J1740.5−3014 (Uchiyama 2010), and d = IGR J17391−3021 (e.g., Bozzo et al. 2010). When extracting the GRXE spectra from the XIS data, these sources are masked. On the other hand, we cannot exclude them in HXD/PIN because it does not have imaging capability. Therefore, the HXD/PIN spectra contain signals from these point sources, in addition to those from the GRXE.

Table 6.1 *Suzaku* observations of the Galactic center analyzed in the present study

	Obs. ID[a]	Coordinate[b]		Start time	Exposure[c]		Count rate[d]	
		l	*b*	UT	XIS	PIN	XIS	PIN
Region 1								
1	501053010	−1.83	−0.00	2006-10-10 21:18:59	21.9	19.9	0.49	0.07
2	503014010	−2.10	−0.05	2008-09-18 04:46:49	55.4	51.2	0.36	0.05
3	503015010	−2.35	−0.05	2008-09-19 07:33:05	56.8	52.8	0.38	0.04
4	503016010	−2.60	−0.05	2008-09-22 06:47:49	52.2	49.3	0.36	0.03
5	503017010	−2.85	−0.05	2008-09-23 08:08:10	51.3	48.6	0.33	0.04
6	503021010	−1.62	0.20	2008-10-04 03:44:03	53.8	49.6	0.49	0.07
7	503076010	−1.50	0.15	2009-02-24 17:04:51	52.9	43.8	0.55	0.07
8	503077010	−1.70	0.14	2009-02-26 01:01:00	51.3	43.7	0.48	0.07
9	504001010	−1.47	−0.26	2010-02-26 09:15:00	51.2	42.2	0.41	0.05
10	504002010	−1.53	−0.58	2010-02-27 16:14:41	53.1	46.6	0.35	0.04
11	504003010	−1.45	−0.87	2010-02-25 04:33:17	50.9	41.3	0.39	0.02
12	504090010	−1.49	−1.18	2009-10-13 04:17:20	41.3	35.0	0.40	0.03
				Total exposure	592.1	524.0		
Region 2								
1	502004010	0.17	−1.00	2007-10-10 15:21:17	19.9	18.8	0.90	0.05
2	502059010	−0.00	−2.00	2007-09-29 01:40:51	136.8	110.5	0.70	0.02
3	503081010	0.03	−1.66	2009-03-09 15:41:50	59.2	57.6	0.98	0.01
4	504050010	0.10	−1.42	2010-03-06 03:55:37	100.4	80.5	1.21	0.02
5	504088010	−0.00	−0.83	2009-10-14 11:30:56	47.2	32.6	0.87	0.05
6	504089010	−0.05	−1.20	2009-10-09 04:05:59	55.3	40.2	1.08	0.02
				Total exposure	418.8	340.2		

[a]Observation ID
[b]Aim point in the Galactic coordinate (degree)
[c]Net exposure in units of 10^3 s
[d]Count rates in units of counts s^{-1} calculated over 1–9 keV (XIS-FI i.e. XIS0 and XIS3) and 17–50 keV (PIN). Particle (non X-ray) background is subtracted from the XIS and the PIN data. The XIS count rate includes CXB which accounts about 0.04 counts s^{-1} whereas it is subtracted in the PIN rates (Sect. 5.1)

We extracted spectra of the point sources from the XIS data of Obs. IDs 504001010, 503021010, and 402066010, and show them in Fig. 6.2. To estimate their fluxes in the HXD/PIN energy band, we fitted the spectra with phenomenological models. Because there is no characteristic feature in the spectra of Sources a, b, and d, we applied to them power-law models subject to the interstellar absorption (i.e., `wabs` × `powerlaw` in XSPEC). Source c shows emission-line like features at 6–7 keV, and is identified as an intermediate polar (IP) by Uchiyama (2010) based on its periodic time variation. Therefore, this source should be fitted with a multi-temperature model (Sect. 5.2.2). Although the counting statistics are not so high, instead of using the IP model (Chap. 3), we restricted ourselves to a thermal plasma model consisting of two collisional ionization equilibrium (CIE) emission components plus a Gaussian subject to photo absorptions [i.e., `wabs` × (`APEC`+`Gaussian`)+`wabs` × `APEC`)]. The two plasma components,

Table 6.2 X-ray point sources in the Galactic center region that are listed in the *INTEGRAL* General Reference Catalog (Ebisawa et al. 2003) version 31

ID	Name	Coordinate[a] (*l*,*b*)	ID	Name	Coordinate[a] (*l*,*b*)
1273	GRS 1730−312	(−3.33, 1.01)	1356	Swift J174540−2900	(−0.05, −0.04)
1279	AX J1735.1−2930	(−1.70, 1.64)	1357	IGR J17456−2901	(−0.04, −0.05)
1283	XB 1732−304	(−2.44, 0.99)	1358	1E 1742.5−2845	(0.14, 0.06)
1285	2CG 356+00	(−3.44, 0.30)	1359	RX J1745.7−2904	(−0.11, −0.10)
1288	AX J1736.4−3141	(−3.38, 0.22)	1360	1E 1742.7−2902	(−0.07, −0.11)
1289	AX J1736.4−2910	(−1.26, 1.57)	1361	1E 1742.8−2853	(0.06, −0.06)
1291	2EGS J1736−2904	(−1.14, 1.51)	1363	1E 1742.9−2852	(0.09, −0.06)
1292	GRS 1734−292	(−1.11, 1.41)	1364	1A 1742−294	(−0.44, −0.39)
1293	IGR J17375−3022	(−2.16, 0.72)	1365	IGR J17461−2853	(0.10, −0.06)
1296	AX J1738.4−2902	(−0.91, 1.28)	1367	1E 1742.9−2849	(0.13, −0.06)
1297	GX 359+2	(−0.43, 1.55)	1369	IGR J17464−3213	(−2.74, −1.83)
1299	IGR J17391−3021	(−1.93, 0.45)	1370	1E 1743.1−2852	(0.12, −0.11)
1300	AX J1739.3−2923	(−1.12, 0.93)	1371	1E 1743.1−2843	(0.26, −0.03)
1301	IGR J17394−3007	(−1.71, 0.52)	1373	IGR J17467−2848	(0.24, −0.14)
1303	AX J1739.5−2910	(−0.90, 1.01)	1374	IGR J17468−2902	(0.04, −0.27)
1304	GRS 1736−297	(−1.37, 0.71)	1376	AX J1747.0−2828	(0.55, −0.03)
1306	XTE J1739−285	(−0.29, 1.30)	1377	AX J1747.0−2837	(0.43, −0.10)
1307	GRS 1737−310	(−2.38, −0.04)	1378	1A 1743−288	(0.21, −0.24)
1308	AX J1740.1−3102	(−2.41, −0.10)	1379	AX J1747.1−2809	(0.84, 0.13)
1309	AX J1740.2−2848	(−0.50, 1.09)	1381	IGR J17464−2811	(0.83, 0.08)
1310	AX J1740.2−2903	(−0.72, 0.92)	1382	IGR J17473−2721	(1.55, 0.51)
1311	AX J1740.4−2856	(−0.61, 0.97)	1383	IGR J17475−2822	(0.65, −0.04)
1313	AX J1740.5−3014	(−1.68, 0.28)	1384	AX J1747.3−2809	(0.87, 0.07)
1314	AX J1740.5−2937	(−1.17, 0.58)	1385	SLX 1744−300	(−0.74, −0.91)
1316	SLX 1737−282	(−0.03, 1.26)	1386	SLX 1744−299	(−0.72, −0.90)
1317	IGR J17407−2808	(0.12, 1.35)	1388	2CG 359−00	(−1.23, −1.30)

(continued)

Table 6.2 (continued)

ID	Name	Coordinate[a] (l,b)	ID	Name	Coordinate[a] (l,b)
1319	EXS 1737.9−2952	(−1.32, 0.34)	1389	IGR J17479−2807	(0.96, −0.01)
1323	IGR J17419−2802	(0.35, 1.16)	1390	GX 3+1	(2.29, 0.79)
1326	AX J1742.5−2845	(−0.20, 0.68)	1392	XTE J1748−288	(0.68, −0.22)
1328	GRS 1739−278	(0.64, 1.17)	1395	AX J1748.3−2854	(0.33, −0.49)
1329	AX J1742.6−3022	(−1.56, −0.20)	1396	AX J1748.6−2957	(−0.54, −1.08)
1330	AX J1742.6−2901	(−0.41, 0.50)	1399	IGR J17487−3124	(−1.75, −1.83)
1332	KS 1739−304	(−1.67, −0.29)	1400	AX J1748.7−2709	(1.88, 0.33)
1335	GC X−4	(−0.63, 0.08)	1404	AX J1749.1−2733	(1.59, 0.05)
1336	1E 1740.7−2942	(−0.88, −0.11)	1405	AX J1749.2−2725	(1.70, 0.12)
1337	AX J1743.9−2846	(−0.06, 0.40)	1406	GRO J1750−27	(2.37, 0.51)
1339	Ginga 1741.2−2859	(−0.20, 0.18)	1407	IGR J17497−2821	(0.95, −0.45)
1340	1E 1742.5−2859	(−0.19, 0.18)	1411	SAX J1750.8−2900	(0.45, −0.95)
1341	IGR J17445−2747	(0.87, 0.83)	1413	IGR J17507−2647	(2.45, 0.18)
1342	GRO J1744−28	(0.05, 0.30)	1414	IGR J17507−2856	(0.57, −0.96)
1343	IGR J17448−3232	(−3.19, −1.70)	1415	GRS 1747−312	(−1.44, −2.17)
1344	SAX J1744.7−2916	(−0.40, −0.01)	1416	AX J1751.1−2748	(1.59, −0.45)
1345	RX J1744.7−2713	(1.36, 1.05)	1417	IGR J17511−3057	(−1.12, −2.07)
1346	KS 1741−293	(−0.43, −0.09)	1419	XTE J1751−305	(−0.82, −1.92)
1348	GRS 1741.9−2853	(−0.05, 0.12)	1425	GX 1.1−1.0	(1.12, −1.03)
1349	IGR J17453−2853	(−0.05, 0.12)	1431	AX J1753.5−2745	(1.91, −0.89)
1350	IGR J17454−2703	(1.58, 1.05)	1433	AX J1754.0−2929	(0.47, −1.85)
1351	1E 1742.2−2857	(−0.06, 0.01)	1436	AX J1754.2−2754	(1.85, −1.10)
1352	GPS 1742−326	(−3.22, −1.93)	1438	AX J1754.5−2927	(0.54, −1.94)
1353	1A 1742−289	(−0.07, −0.04)	1441	AX J1755.7−2818	(1.68, −1.59)
1354	AX J1745.6−2900	(−0.07, −0.05)	1443	AX J1758.0−2818	(1.94, −2.02)
1355	IGR J17456−2901b	(−0.08, −0.05)			

[a] Source position in the Galactic coordinate (degree)

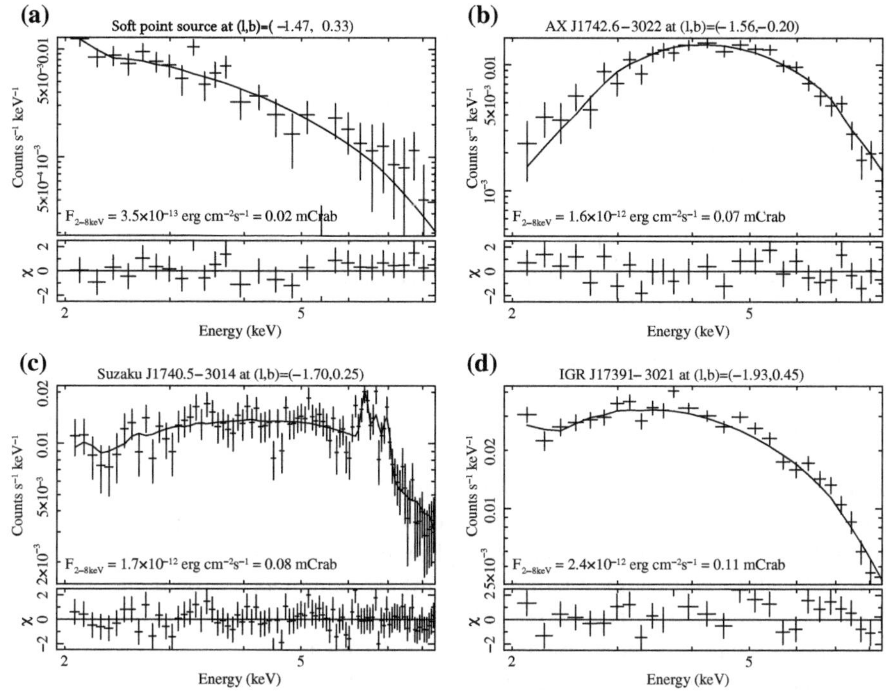

Fig. 6.2 Background-subtracted XIS spectra of four contaminating sources recognized in Region 1; **a** a newly-found soft X-ray point source, **b** AX J1742.6−3022, **c** Suzaku J1740.5−3014, and **d** IGR J17391−3021. *Crosses* and *solid curves* in *upper panels* are data and the best-fit models, respectively. *Lower panels* present fitting residuals in terms of χ. *Inset labels* represent model-predicted 2–8 keV energy fluxes from the sources, followed by the same value scaled by that of the Crab nebula

and the Gaussian represent typical high and low temperatures of a multi-temperature plasma in an accretion column, and reprocessed emission line from neutral Fe, respectively.

In all cases, the fits successfully reproduced the observed spectra of these sources (Fig. 6.2), and provided best-fit parameters as listed in Table 6.3. Model-predicted intensities in the 2–8 keV band were <0.11 mCrab for all the sources. When HXD/PIN observes these sources in its field of view, the instruments angular transmission (Sect. 4.1.2) reduces these fluxes, and therefore their effective contributions become much smaller. Sources a and b have rather soft photon indices compared to that of the Crab nebula ($\Gamma = 2.1$), and hence their intensities in the HXD/PIN energy band further decrease when expressed in units of Crab (or mCrab).

From these evaluations, we conclude that the total contribution from these point sources to the HXD/PIN data is far less than 0.1 mCrab, or equivalently 3×10^{-3} counts s^{-1} in HXD/PIN over the 15–50 keV band. This means that the present analysis integrates X-ray signals emitted from unresolved sources fainter

Table 6.3 The best-fit parameters for the contaminating sources

	Source	Model[a]	$n_{\rm H}$ ($\rm cm^{-2}$)	$\Gamma^{\rm b}$	$kT^{\rm c}$ (keV)	χ^2/ν	$F^{\rm d}_{2-8\ \rm keV}$ (mCrab)
a	New source	PL	$2.4^{+3.7}_{-1.9}$	$3.1^{+1.5}_{-0.7}$	–	1.20(25)	0.01
b	AX J1742.6−3022	PL	13^{+2}_{-2}	$2.7^{+0.3}_{-0.4}$	–	0.91(28)	0.07
c	Suzaku J1740.5−3014	2CIE	$3.7^{+1.0}_{-1.2}$	–	10.8(>6.5)	0.81(86)	0.07
			45^{+78}_{-35}	–	31(> 4)		
d	IGR J17391−3021	PL	$4.0^{+0.8}_{-0.6}$	$1.9^{+0.2}_{-0.1}$	–	1.32(25)	0.11

[a] *PL* an absorbed power-law model, *2CIE* two CIE plasma plus a Gaussian model
[b] Photon index of the power-law model
[c] Plasma temperature. Numbers in brackets are lower limits on the 90 % confidence level
[d] Model predicted flux in the 2–8 keV band (1Crab = 2.2×10^{-8} erg cm^{-2} s^{-1})

than 0.1 mCrab, or $\sim 2 \times 10^{12}$ erg cm^{-2} s^{-1} in the 2–8 keV band as the "GRXE". This limiting flux corresponds to an intrinsic X-ray luminosity of a point source of 8×10^{33} erg s^{-1} if a distance to the Galactic center of 8 kpc is assumed. This limit is several (at most) to ten times lower the previous studies (Revnivtsev et al. 2006; Türler et al. 2010), thanks to the tightly-collimated field of view of the HXD ($34' \times 34'$ at the full width at half maximum) and simultaneous imaging capability of the XIS images. Since typical HXD/PIN count rates in Regions 1 and 2 are ~0.04 and ~0.02 counts s^{-1}, contamination from $\lesssim$0.1 mCrab point source corresponds to less than a few to ten percent of the detected flux, and is negligible, if any exists, compared to the NXB subtraction accuracy.

A question might be raised about possible existence of faint point sources outside the regions covered by the XIS. The *ASCA* survey around this region (Sakano et al. 2002) can be a nice reference for investigating such sources because it is the only sensitive imaging survey in this sky region, and in the X-ray energy band available at this moment. The survey covers $5° \times 5°$ around the Galactic center, and have a point source sensitivity of ~several$\times 10^{-13}$ erg cm^{-2} s^{-1} in the 0.7–10 keV (equivalently ~0.05 mCrab). As can be recognized from the prefix AX (meaning *ASCA* X-ray sources) in Table 6.2, point sources detected in the *ASCA* survey are already included in the *INTEGRAL* catalog, and sources could fall inside the HXD/PIN field of view (IDs 1304, 1307, 1308, 1314, and 1288) are all very faint ($\ll$0.1 mCrab), and they do not affect our study at all.

Among the four sources considered here, IGR J17391−3021 is known as a typical Super-giant Fast X-ray Transient source which is characterized by a low intrinsic luminosity during quiescence and sporadic short time X-ray flares lasting 10^3–10^4 s. Although flare activities involving count increase by 10^3–10^4 times from this source have been reported (e.g., Sakano et al. 2002; Blay et al. 2008; Sidoli et al. 2009), the present data containing this source inside the HXD/PIN field of view did not show variability beyond the Poisson distribution (see the next section). Therefore, we consider that the source was in a quiescent state during our observations, and hence, the above arguments on the contaminating point sources remain valid.

6.1.3 Time Variations

Regardless of the origin of the GRXE, time variation is not expected; largely extended diffuse plasma whose size is hundreds to thousand parsec cannot vary with a timescale of days ("Diffuse" scenario), and it is hardly possible that numerous faint X-ray point sources synchronously change their luminosity ("Point Source" scenario). However, if any bright point source is inside the field of view of the selected observations, and if it varies on a timescale of days, we should see count rate variations during one observation. Therefore it is also important, before performing detailed spectral analysis, to confirm that the background-subtracted signals do not exhibit significant time variations.

The above issue applies particularly to the HXD/PIN, because it lacks imaging capability, and the outer 75 % (or 50 %, after considering angular transmission) of its field of view falls outside that of the XIS. To examine if such sources were inside the HXD/PIN field of view and actively emitting hard X-rays, we created HXD/PIN light curves of each observation. Figures 6.3 and 6.4 show the results from four representative observations; Obs. IDs 503017010 and 504002010 from Region 1, and

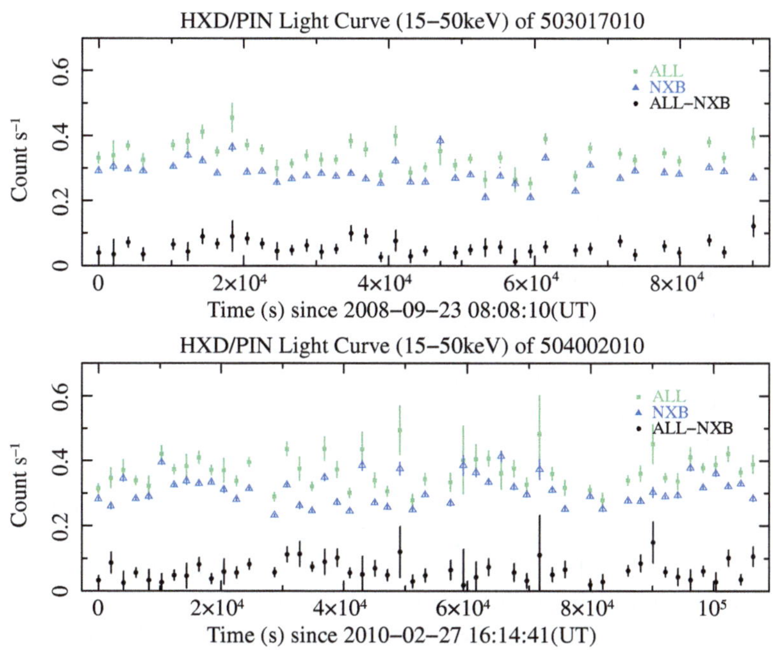

Fig. 6.3 Examples of light curves of the HXD/PIN count rates during the GRXE observations, integrated over the 15–50 keV band; Obs. IDs 503017010 and 504002010 from Region 1. ALL (*green filled rectangles*), NXB (*blue open triangles*), and ALL-NXB (*black filled circles*) are raw detector count rates, simulated count rates of the NXB, and NXB-subtracted celestial signals including the CXB. Each bin has a width of 2000 s

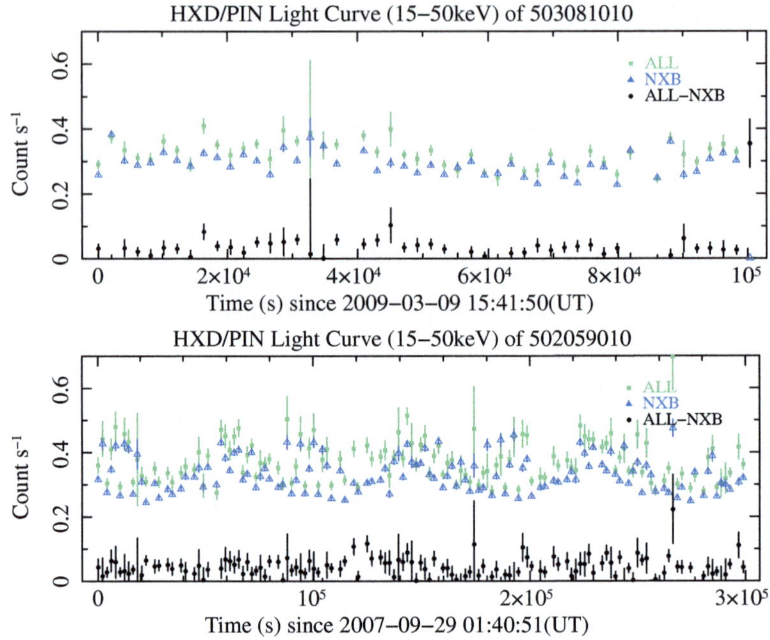

Fig. 6.4 The same as Fig. 6.3, but for two examples from Region 2 observations (Obs. ID 503081010 and 502059010)

503081010 and 502059010 from Region 2. After subtracting the NXB, time-averaged residual count rates in these observations were ~0.02–0.07 counts s^{-1}. Since the CXB amounts to ~0.015 counts s^{-1} in HXD/PIN, about 50–70 % of the residual counts can be considered to originate from the GRXE. Periodic time variations of the NXB count rate, especially apparent in Obs. ID 502059010, is caused basically by the change of the cosmic-ray flux as the satellite orbits. Long time gap between data points occasionally taking place in the light curves are due to chance coincidence of the Earth occultation and a passage of the South Atlantic Anomaly in which the instrument is intentionally stopped (Sect. 4.1) due to the very high flux of cosmic-ray particles.

We examined possible temporal variations by fitting the light curves with a zero-th order function (constant). In all cases, the derived reduced chi-squares were ~1. Therefore, the present HXD/PIN signals are consistent with being constant during each observation, without intrinsic variation beyond the Poisson fluctuations. Based on this, we conclude that none of bright transient point sources contaminates our HXD/PIN observations. We, then, proceed to spectral analysis using the data obtained in these observations.

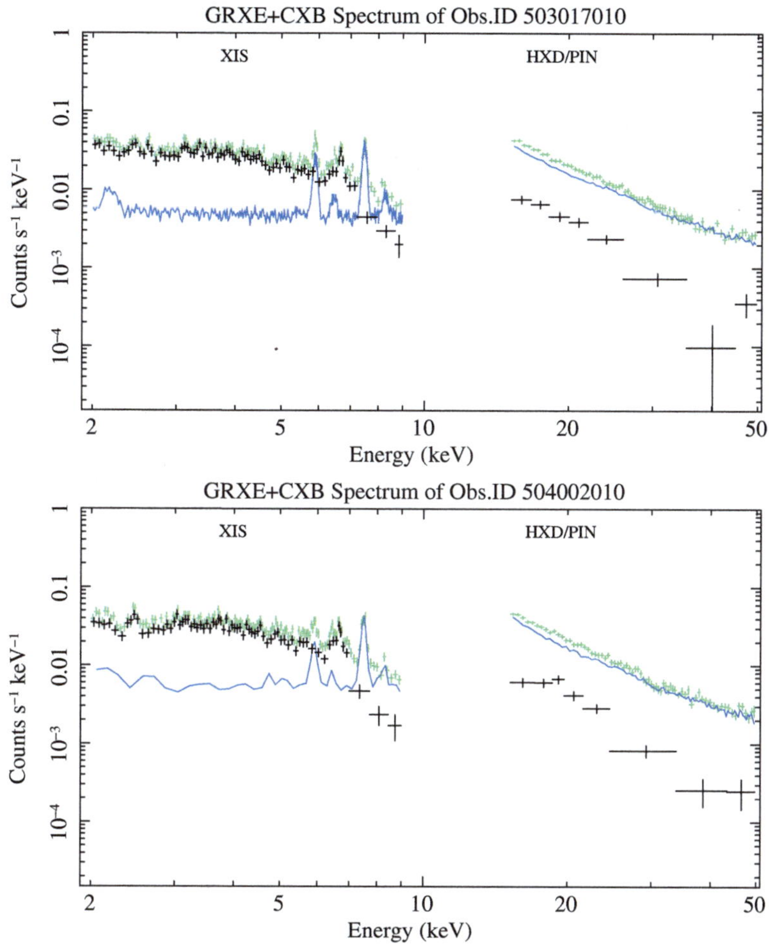

Fig. 6.5 Examples of wide-band X-ray spectra obtained in individual observations in Region 1. *Black crosses* are the celestial signal counts, which consist of the GRXE and the CXB, derived by subtracting the NXB (*blue solid line*) from the raw counts (*green crosses*). For clarity, spectra of XIS0 are plotted, although the other two XIS cameras which give fully consistent data

6.1.4 Individual Spectra

We extracted X-ray spectra from individual observations. The XIS data of all the imaging area of the CCD were used, except for two corners irradiated by calibration radioactive isotopes (Fig. 4.5). Figures 6.5 and 6.6 show examples of the spectra, each derived from the same observations as were used in the previous light curve analysis. The spectra of the raw data are shown as "ALL" together with the "NXB" and "ALL-NXB" (NXB-subtracted celestial signals) spectra. Note that since the "ALL-NXB" contains the signals from not only the GRXE but those from the CXB which

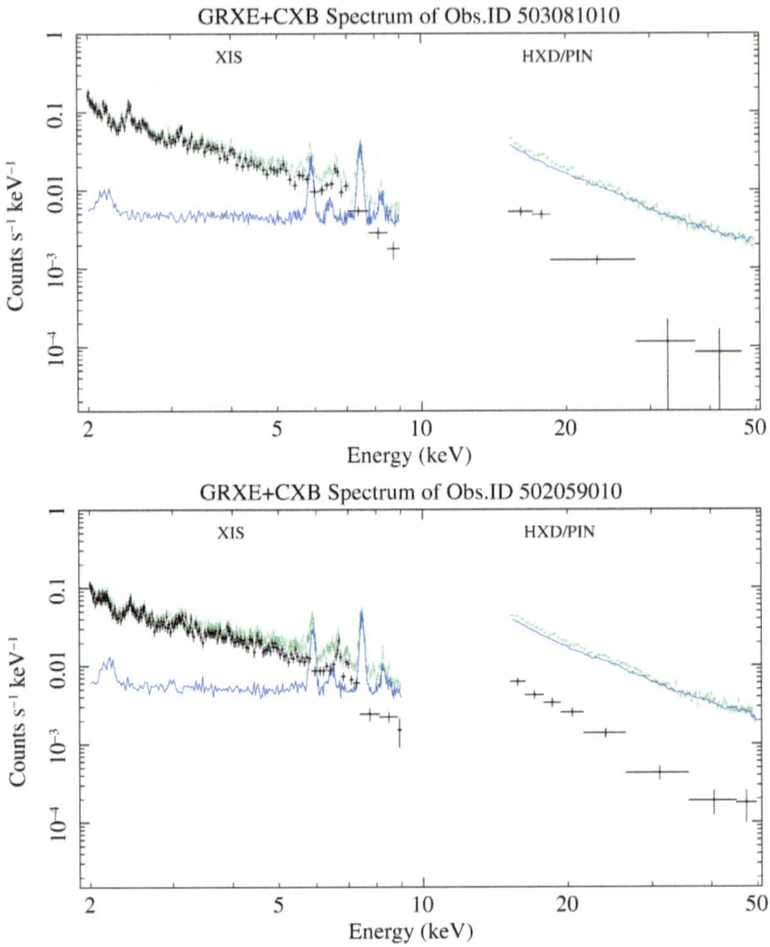

Fig. 6.6 The same as Fig. 6.5, but obtained from two typical observations from Region 2

should be taken into account in fit models. Here, we tentatively used energy ranges of 2–9 keV and 15–50 keV for the XIS and HXD/PIN, respectively. These ranges are optimized again in Sect. 6.1.5 for a summed wide-band spectrum. The XIS NXB was estimated by applying the standard tool `xisnxbgen` to the same extracting region as the data. The same NXB model as used in the IP analysis (Sect. 5.1.2) was used for subtracting HXD/PIN background.

As can be expected from the count rates in the light curves, the NXB subtraction has left signals which are significant compared to the statistical and systematic uncertainties of the NXB subtraction. We regard this as the GRXE signal, again with the 30–50 % contribution from the CXB. In the HXD/PIN energy band, however, the signal count is fairly low, reflecting the low surface brightness of the GRXE.

For detailed wide-band spectral analyses, we need to combine the data of several observations.

The intense Fe Kα emission lines (6–7 keV) are noticeable even in the individual XIS spectra. A difference of spectral slopes below ~3–4 keV between the two regions are caused by the stronger interstellar absorption on the Galactic plane (Region 1) than that in the off-plane regions (Region 2). The origin of this is discussed again based on spectral fitting results (Sect. 6.2.3). In addition, several emission-line like features which are thought to originate from lighter elements are recognizable in 2–4 keV (e.g., at around 2.5 and 3.2 keV). In Sect. 6.2.2, we study these features in detail, and identify their origin.

6.1.5 Wide-Band Spectra

Apart from the systematic inter-region difference, the spectra are very similar to one another among each Region (1 or 2). Therefore, we summed the spectra of the GRXE (and the CXB) over all the observations in each Region. Resulting spectra are shown in Fig. 6.7. Based on the NXB estimation accuracies, ~5 % in the XIS (Tawa et al. 2008) and ~1 % in HXD/PIN (Fukazawa et al. 2009), we consider that the celestial signals are detected up to 10 keV and 50 keV in the XIS and HXD/PIN. Compared to the individual spectra, the emission lines from Fe and other lighter elements can be more clearly seen in the XIS spectra. For better visibility, we present a close-up view of the Fe emission lines in the 6–7 keV band in Fig. 6.8.

Although the XIS detects significant signals below 2 keV as well, we did not use them because detailed spectral modeling in the energy band is hampered by interstellar absorption. In addition, this energy region is contributed by an additional soft component, of which the origin is controversial, either foreground diffuse plasma emission or the solar wind charge exchange (Kuntz and Snowden 2008; Yoshino et al. 2009; Masui et al. 2009), neither of which is relevant to the GRXE. The HXD/PIN spectra below 15 keV were not used because of enhanced instrumental noise caused by the radiation damage of the PIN detectors and occasional high operation temperature of the HXD, which was not accurately reproduced by the NXB model. In Region 2, we limited the spectra to narrower energy ranges (2–9 keV for the XIS and 15–40 keV for HXD/PIN) due to lower counting statistics because of the shorter exposure and the lower surface brightness of the GRXE in the off-plane region.

6.1.6 Surface Brightness Model Based on the Near IR Diffuse Emission

For spectral analyses, energy responses and effective areas of the XIS and HXD/PIN should be calculated for individual observations by taking into account a global distribution of the GRXE surface brightness. The GRXE surface brightness is known to correlate with that of near infrared diffuse emission which reflects the stellar mass

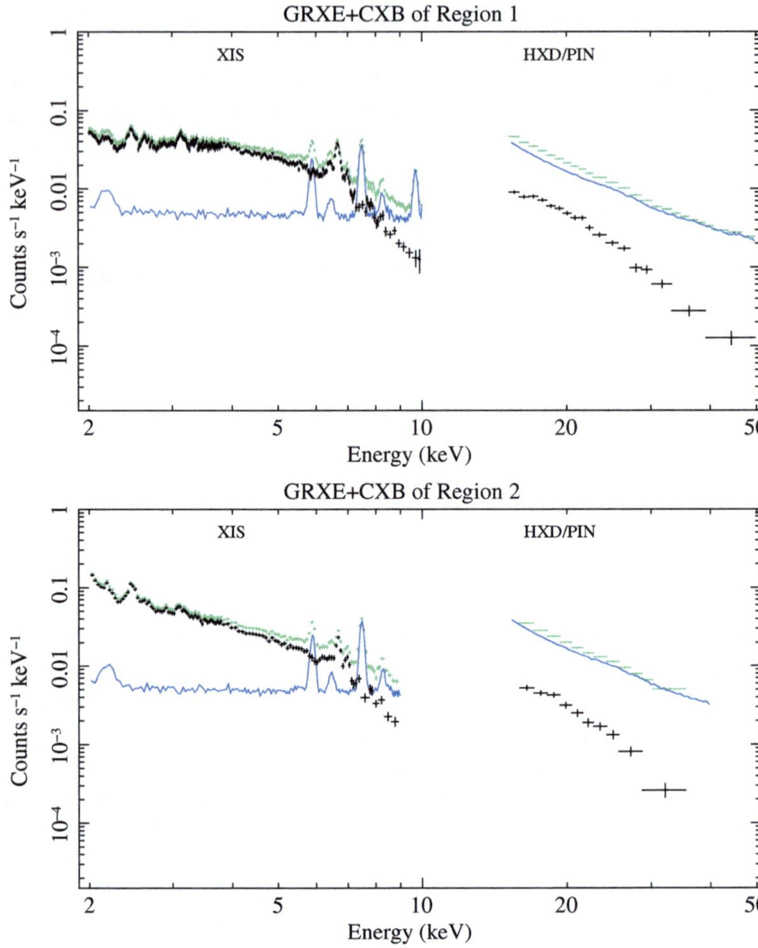

Fig. 6.7 Spectra of the GRXE (including the CXB) summed over all observations in Region 1 (*upper panel*) and Region 2 (*lower panel*), presented in the same way as Fig. 6.5

density, i.e., the density of stars (Sect. 2.1.1). Among several functional models, one called G3 (Dwek et al. 1995) was used to fit the GRXE surface brightness in Revnivtsev et al. (2006). The model assumes a stellar density $\rho(r)$ as

$$\rho(x, y, z) = \rho_0 r^{-1.8} \exp(-r^3), \tag{6.1}$$

$$r = \left[\left(\frac{x}{x_0}\right)^2 + \left(\frac{y}{y_0}\right)^2 + \left(\frac{z}{z_0}\right)^2\right]^{1/2}, \tag{6.2}$$

where x, y, and z are three-dimensional coordinates centered on the Galactic center with the z axis being aligned with the rotation axis of the Galaxy and x_0, y_0, and z_0

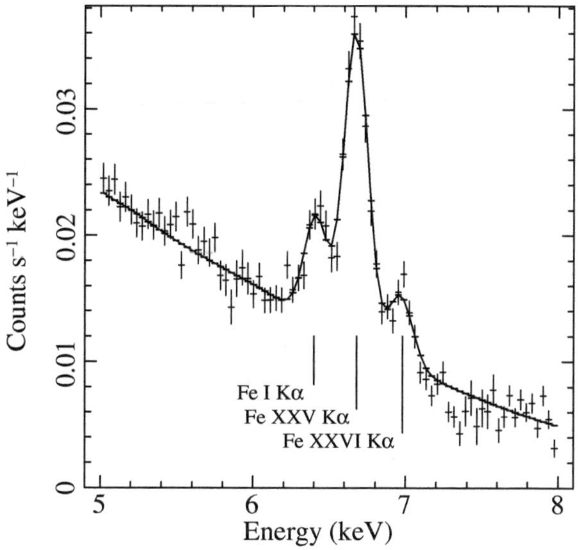

Fig. 6.8 *Close-up view* of the Fe emission lines of the *upper panel* in Fig. 6.7. Three Kα lines from Fe in different ionization states are clearly resolved; Fe I = neutral, Fe XXV = He-like, and Fe XXVI = H-like

are scale-length parameters; Dwek et al. (1995) gave $x_0 = 4.01$ kpc, $y_0 = 1.67$ kpc, and $z_0 = 1.12$ kpc. For detailed geometrical configuration including rotation around z axis, see Dwek et al. (1995). The surface brightness S at a certain sky position was assumed to follow a line-of-sight integral,

$$S(l, b) = S_0 \int_0^{\infty} \rho(x, y, z) \mathrm{d}s, \tag{6.3}$$

where (l, b) is a sky position in the Galactic coordinate, and S_0 and s are a normalization parameter, and a line-of-sight distance (x, y, z) measured from the Sun, respectively. The Galactocentric distance to the Sun was assumed to be 8.5 kpc in this calculation.

Based on this, we constructed the surface brightness map as presented in Fig. 6.9, and used this as an input in calculating the detector responses. To justify the procedure, we compared the actual measured HXD/PIN count rate of each observation with the near-infrared surface brightness of Eq. (6.3) convolved through the HXD/PIN angular response. We chose HXD/PIN because the interstellar absorption does not affect the count rate unlike below 10 keV, and the CXB count rate can be accurately subtracted (while this is not possible in the XIS). A nice correlation was obtained as plotted in Fig. 6.10. We also confirmed a similar (but less tight due to the position-dependent absorption) correlation in the XIS count rates. After the individual calculation, we added the responses by weighting them with their respective exposures.

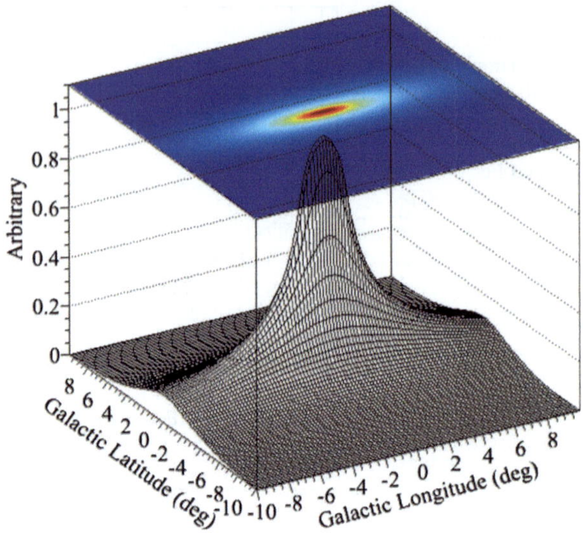

Fig. 6.9 A surface brightness model of the GRXE which we calculated using the stellar density model of Dwek et al. (1995)

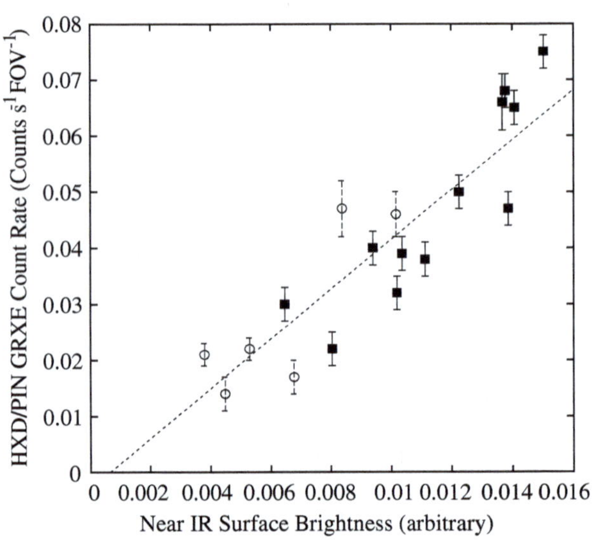

Fig. 6.10 Correlation of the HXD/PIN GRXE count rate (15–50 keV) and the near IR surface brightness (Dwek et al. 1995) convolved with the HXD/PIN angular transmission ($34' \times 34'$ at the full width at half maximum). Count rates of individual observations of Regions 1 and 2 are plotted in *filled rectangles* and *open circles*, respectively. The *dashed line* is the best-fit linear function ($y = 4.4x - 0.003$). The CXB count rate of 0.016 counts s^{-1} was also subtracted in the HXD/PIN count rates

6.2 X-ray Spectral Analysis

As discussed in the previous section, the spectra, or portions of them, are considered to arise from the GRXE. The fact that contamination from apparently bright point X-ray sources is very little allows us to investigate wide-band spectral properties of the GRXE. We first concentrate on the hard X-ray spectrum taken with HXD/PIN in Sect. 6.2.1 because there is only one type of X-ray source proposed as a candidate of the origin, namely magnetic CVs. Then, using the XIS spectrum, we show that multi-temperature nature of the GRXE (Sect. 6.2.2), and based on this knowledge, we construct a model to decompose the GRXE spectrum into contributions from several types of faint X-ray point sources (Sect. 6.2.3).

6.2.1 Characterization of the Hard X-ray Spectra

To characterize the hard X-ray spectrum of the GRXE, we apply three different models in the following subsections; (1) the IP spectral mode, (2) a single temperature plasma emission mode, and (3) a simple power-law model. Then, we consider the first one most appropriate.

6.2.1.1 (1) The IP Model

Magnetic CVs, especially IPs, are thought to mainly contribute to the GRXE in the hard X-ray band (Sect. 2.1.6; Revnivtsev et al. 2008; Sazonov et al. 2006). X-ray spectral shapes of IPs and polars are quite similar, both having the thermal nature, with the only difference being that the plasma temperatures of polars are lower due to enhanced cyclotron cooling. Therefore, the GRXE spectra integrated above 15 keV should be approximately reproduced by the IP spectral model (Chap. 3).

To test this idea, we fitted the HXD/PIN spectra of Regions 1 and 2 with the IP model. The models for Regions 1 and 2 shared the WD mass parameter, and allowed to have different normalization. The Fe abundance parameter was fixed at unity in the fitting because it cannot be constrained without data of the emission lines. Even when we changed the parameter, for example to 0.5 $Z_{\odot}$, result was not affected at all. The best-fit model successfully reproduced the overall spectral shape in the two regions, and gave $\chi^2_{\nu} = 1.37(30)$ and a null hypothesis probability of 9 %. Therefore, we consider that this fit is acceptable. Figure 6.11 shows the spectra overlaid with the best-fit model. The best-fit WD mass parameter is $M_{\rm WD} = 0.66^{+0.09}_{-0.07}\ M_{\odot}$. In the following analyses, we simply use this fact that the hard X-ray spectra of Regions 1 and 2 can be nicely reproduced with the IP model of $M_{\rm WD} = 0.66\ M_{\odot}$, to fix the parameter when constructing and fitting a wide-band spectral model.

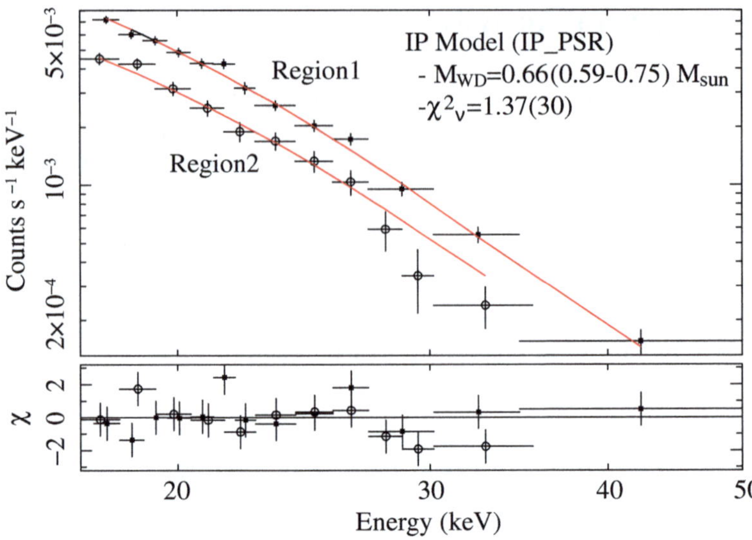

Fig. 6.11 HXD/PIN spectra of the GRXE obtained from Region 1 (*filled rectangles*) and Region 2 (*open circles*) simultaneously fitted with the IP spectral model. The *red solid curves* are the best-fit model spectra (CXB inclusive). The *crosses* in the *lower panel* show fitting residuals

6.2.1.2 (2) A Single-Temperature CIE Plasma Model

Following the successful fit with the IP model, we also tried to fit the spectra with a model of a single-temperature plasma in the CIE (the APEC model in XSPEC). This is a simplified approximation to the IP model because it is expressed as the sum of many APEC components (Sect. 3.5), and the temperature to be defined here can be interpreted as a value representing such distributed temperatures.

By this procedure, an acceptable fit with $\chi^2_\nu = 1.37(31)$ was obtained, and the best-fit plasma temperature was $kT = 15.7^{+2.7}_{-2.0}$ keV. If we simply take this value as a shock temperature and convert it to a WD mass using the kT_s-M_{WD} relation (the shock temperature and the WD mass relation) presented in Fig. 3.4, a WD mass of 0.48 $M_\odot$ is derived. The fact that this WD mass is lower than that obtained by the IP model fit (above) is reasonable, because the "color temperature" of the IP model is in fact lower than the actual shock temperature (Sect. 3.5).

6.2.1.3 (3) A Simple Power-Law Model

In previous studies of the GRXE, hard X-ray spectra were phenomenologically fitted with a power-law model. As explained in Sect. 2.1.2, a certain kind of diffuse hard X-ray emission via cosmic-ray induced non-thermal process was a background assumption of this modeling.

For comparison with the above two models, we also tried a power-law fitting to our hard X-ray spectra (Fig. 6.11). In this model, the photon flux f at a certain energy E is defined as $f(E) = f_0(E/1\,\text{keV})^{-\Gamma}$, where f_0 is a normalization parameter, and Γ is a photon index which determines the spectral slope. The best fit gave slightly worse but acceptable fitting statistics $\chi^2_\nu = 1.43(31)$, and derived a power-law index of $\Gamma = 2.8 \pm 0.2$. This photon index is softer when compared with the value derived from the *RXTE* and OSSE fitting (2.3 ± 0.2; Valinia and Marshall 1998).

If this power-law function is extrapolated down to the XIS energy band, the model-predicted count rates of the XIS exceeds the data by more than a factor of two as illustrated in Fig. 6.12. This means that the derived power-law fit is (phenomenologically) valid only in the HXD/PIN energy band, and does not reproduce the spectra of the XIS. To solve this discrepancy of the data and the power-law model, very dense absorption ($n_{\rm H} \sim 2 \times 10^{23}$ cm^{-2}) is necessary to decrease photon flux in the energy band. However, this seems very unrealistic because the required absorption column cannot be explained by an interstellar column density toward the Galactic center, typically several$\times 10^{22}$ cm^{-2}. Thus, although the power-law fit is statistically acceptable, we consider that this is not an appropriate (or physically reasonable) model to interpret the data.

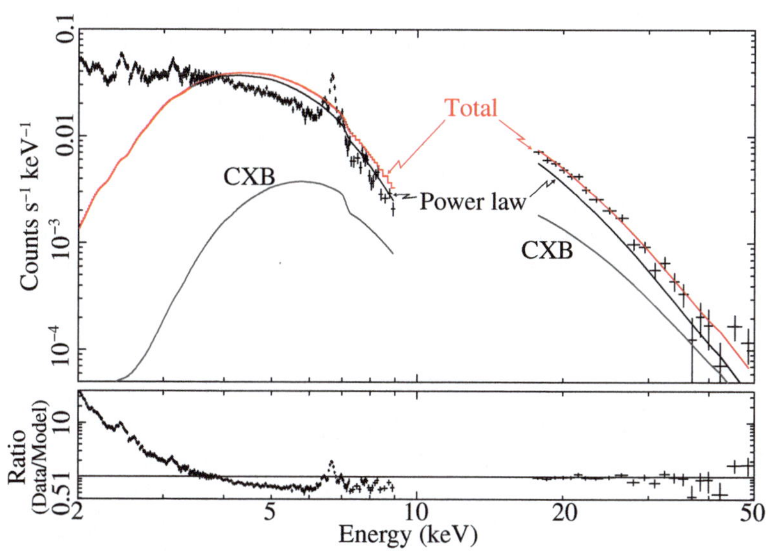

Fig. 6.12 The wide-band GRXE spectra of Region 1 and the power-law model of $\Gamma = 2.8$ (*black curve*) extrapolated from the HXD/PIN (>10 keV) band to the XIS band (<10 keV). An interstellar absorption by a typical column density of $n_{\rm H} = 6 \times 10^{22}$ cm^{-2} is applied to the model. *Red curves* are the sum of the power-law and the CXB (*gray curve*) components

6.2.2 Multi-Temperature Nature of the GRXE

Following the hard X-ray spectral analysis, we then concentrate on the soft X-ray spectrum (2–10 keV) of the GRXE taken with the XIS. For simplicity, we analyze the Region 1 spectrum of XIS0 only. The obtained results are not changed if we additionally utilize the spectra of XIS1 and XIS3. The same discussion can be applied to the Region 2 spectrum, with a natural modification of the interstellar absorption column density.

6.2.2.1 Identification of Atomic Emission Lines

As can be seen in Fig. 6.7, the XIS spectra shows many emission lines from astrophysically abundant heavy metals. These lines, especially their intensities, provide crucial information on the temperature of a plasma which emits the lines, providing a great help in understanding the GRXE.

To identify atoms and their ionization states from which these lines are emitted, we fitted the XIS spectrum with a phenomenological model consisting of a power-law continuum subject to an interstellar absorption, and a certain number of Gaussians. As shown in Fig. 6.13, the model successfully reproduced the spectrum when we introduced eight Gaussians for prominent line features. As listed in Table 6.4, the center energies of the Gaussians are, from lower to higher, 2.44, 2.62, 3.13, 3.34, and 3.87 keV, with a typical fitting error of 10–70 eV. With these center energies, identification is rather straightforward, and we found that He-like or H-like S, Ar, Ca, and Fe ions are emitting these lines as labeled in the figure.

In the 2–3 keV energy range, the foreground diffuse X-ray emission is known to contribute to some extent to the emission lines (e.g., Kuntz and Snowden 2008; Yoshino et al. 2009; Masui et al. 2009). However, the energy flux attributable to

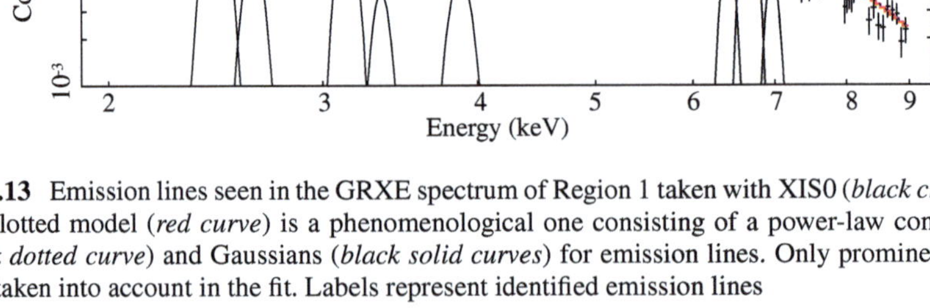

Fig. 6.13 Emission lines seen in the GRXE spectrum of Region 1 taken with XIS0 (*black crosses*). The plotted model (*red curve*) is a phenomenological one consisting of a power-law continuum (*black dotted curve*) and Gaussians (*black solid curves*) for emission lines. Only prominent lines were taken into account in the fit. Labels represent identified emission lines

Table 6.4 Best-fit parameters for individual emission lines seen in the XIS spectrum of the GRXE

Line	Energy (keV)	Line width[a] (eV)	Intensity[b] (eV)
S XV Kα	2.45 ± 0.01	<35	25^{+2}_{-3}
S XVI Kα	2.63 ± 0.01	<25	35^{+8}_{-9}
Ar XVII Kα	3.13 ± 0.01	<11	41^{+6}_{-6}
Ar XVIII Kα	3.33 ± 0.02	<40	14^{+7}_{-8}
Ca XIX Kα	3.86 ± 0.07	<89	27^{+7}_{-9}
Fe I Kα	6.41 ± 0.01	<36	76^{+9}_{-10}
Fe XXV Kα	6.68 ± 0.01	<17	322^{+14}_{-15}
Fe XXVI Kα	6.97 ± 0.01	<34	79^{+10}_{-11}

[a]Upper limits (90 % confidence level) of the intrinsic line width in terms of 1σ of Gaussian
[b]Line intensities calculated against the continuum flux at the line center energies

this emission is only a few percent of the detected flux of the GRXE, and therefore, we neglect its contribution at this stage. Instead, in a wide-band spectral analysis presented below, we take into account this foreground emission.

In the spectra of Fig. 6.13, the Gaussians incorporated for representing atomic emission lines appear "broad" with finite widths of ~150 eV. However, this is primarily caused by the energy resolution of the CCD used in the XIS (~150–200 eV at the full width at half maximum), and is seen in the spectra because we show the data and the model without removing the instrumental response (Sect. 4.3). In the best-fit result, the widths (σ in Table 6.4) are consistent with zero, yielding upper limits of $\lesssim$10–90 eV at 90 % confidence level. Therefore, they can be considered as narrow lines with no "broadening". Below, this narrow-line result is used to reject some physical emission mechanism proposed as the GRXE origin in previous studies.

6.2.2.2 Fit with a Multi-Temperature Spectral Model

It is almost obvious from the above identification that at least two distinct plasma components compose the GRXE. This is because, as long as collisional ionization equilibrium is assumed, elements like S and Ar would be almost fully ionized (all electrons stripped) and would not emit atomic emission lines in a plasma which is hot enough ($10^{7.5}$–10^8 K) to ionize Fe to H-like to produces Fe XXVI Kα (6.9 keV) line; see the ionization balance calculated by, for example, Bryans et al. (2006, 2009). Note that the multi-temperature nature of the GRXE was first revealed by Kaneda et al. (1997) based on similar arguments on the emission lines from several kinds of elements. The existence of the Fe I (neutral) Kα manifests a reprocessing of X-rays on a cold matter as well. There, the X-rays coming from the hot plasma are photo-absorbed by Fe, and then it exhibits line X-ray emission via electron cascades.

To quantify the multi-temperature nature of the GRXE found above, we first performed a fit to the same spectra with a single-temperature CIE plasma model.

In the fitting, we also included the CXB component which only provides continuum (no emission lines). As shown in Fig. 6.14a, the best-fit model (kT ~4 keV) significantly underpredicts the S XV Kα (2.44 keV) and Ar XVII Kα (3.13 keV) line fluxes, and is unacceptable with $\chi^2_\nu = 2.91(245)$. The deficits seen at S XVI Kα and Ar XVII Kα cannot be recovered even when the abundance parameter is changed because, in a single-temperature CIE plasma, ratios of line fluxes from ions in different ionization states (e.g., S XV Kα/S XVI Kα) are determined simply by the temperature. Thus, the GRXE spectrum cannot be explained by the single-temperature CIE plasma model.

After Kaneda et al. (1997), we introduced an additional lower-temperature CIE component to better reproduce the low-energy lines. The fit improved considerably to $\chi^2_\nu = 1.46(243)$, and best-fit temperatures of $kT_1 = 0.79^{+0.46}_{-0.36}$ keV and $kT_2 = 6.23^{+0.26}_{-0.26}$ keV were obtained. Strictly speaking, this reduced χ^2 is not yet acceptable yielding a null hypothesis probability of less than 1 %. We consider that this is because (1) the XIS instrumental response has low accuracy in the 2–4 keV band due to limited calibration information (Nobukawa et al. 2010) compared to that in the 5–8 keV, and (2) the data (especially line profiles) are not well reproduced by the model convolved with the response.[1] If we include a model systematic error of 2 % to simulate the insufficient response accuracy, the reduced χ^2 decreases to 1.22 with the same degrees of freedom (243), and the fit becomes formally acceptable with a null hypothesis probability of 1 %. Since the obtained best-fit parameters did not change from those of the original fit, we do not investigate this response uncertainty any further in the present study, and consider the parameters derived without the model systematic error as our best estimate with the present data set.

The derived two plasma temperatures are just representative ones, and may not directly reflect actual distributions of the plasma temperatures. For example, the two temperatures may be approximating a mixture of more than two distinct plasma temperatures. Or close, the plasma may even have a continuous temperature distribution over a certain range. However, the data statistically do not require any additional component, and therefore, we do not perform further spectral modeling here. Instead, in the following section, we proceed to wide-band spectral analysis including the HXD/PIN data.

6.2.2.3 On the Fe Kα Emission Lines

We clearly detected and identified the three Fe Kα emission lines, and they were all "narrow" lines without being intrinsically broadened. In some previous studies, data with lower quality (energy resolution and energy scale) gave different Fe emission line shape and width. These led authors to propose exotic scenarios which are simply rejected by the present data.

Kaneda et al. (1997) used a plasma emission model which assumes non equilibrium ionization (NEI). This was because the Fe XXVI (H-like) Kα emission line

[1] This means that a fit would not be acceptable even if a model perfectly reproduces the observed spectral shape.

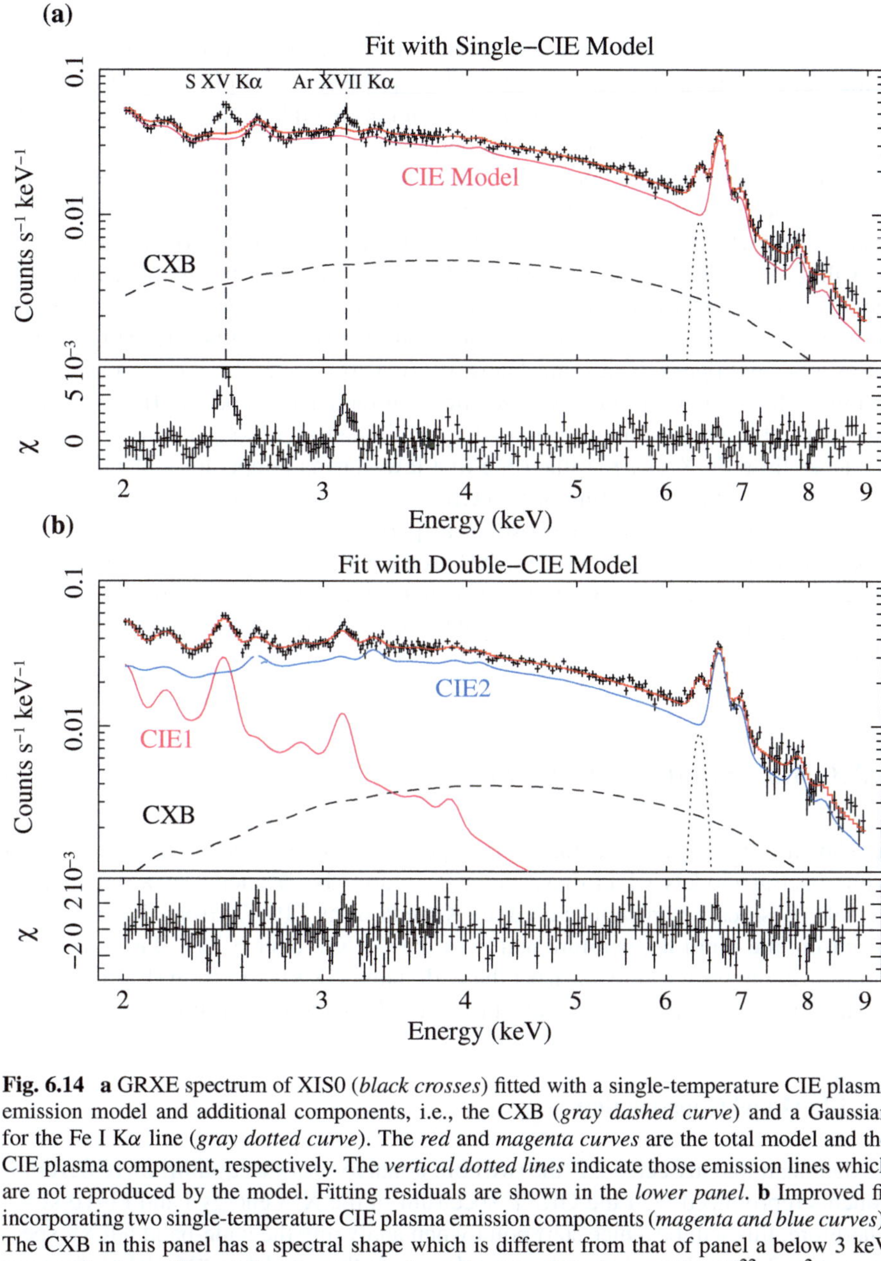

Fig. 6.14 **a** GRXE spectrum of XIS0 (*black crosses*) fitted with a single-temperature CIE plasma emission model and additional components, i.e., the CXB (*gray dashed curve*) and a Gaussian for the Fe I Kα line (*gray dotted curve*). The *red* and *magenta curves* are the total model and the CIE plasma component, respectively. The *vertical dotted lines* indicate those emission lines which are not reproduced by the model. Fitting residuals are shown in the *lower panel*. **b** Improved fit incorporating two single-temperature CIE plasma emission components (*magenta and blue curves*). The CXB in this panel has a spectral shape which is different from that of panel a below 3 keV because the derived interstellar absorption column densities differ ($n_{\rm H} = 2.8 \times 10^{22}$ cm^{-2} in panel **a**, while $n_{\rm H} = 4.8 \times 10^{22}$ cm^{-2} in **b**)

(6.9 keV) seemed absent in their *ASCA* GRXE spectrum although a hard (hot temperature) continuum spectral shape predicted it if an ionization equilibrium was achieved. If ionization is on the way to the equilibrium in a plasma (via Coulomb collision of electrons and ions), the numbers of ions in higher ionization states will be less than those expected from the CIE condition (as can be seen in supernovae). Our present data indicates the intense Fe XXVI Kα line, and therefore, this NEI interpretation lacks observational evidence. Similar arguments can be found in Ebisawa et al. (2008) who also detected the Fe XXVI Kα line in their GRXE spectrum.

Valinia et al. (2000) and Tanaka (2002) also discussed the multi-temperature nature using a CIE plasma model, but at the same time, they included a non-thermal component which is produced by interactions between cosmic-ray particles and interstellar hydrogens. These mechanisms are also rejected by the present data as follows. The existence of Fe XXVI Kα line contradicts the model (CIE plasma plus non-thermal bremsstrahlung) proposed by Valinia et al. (2000) because the model produces little Fe XXVI Kα line (due to a low plasma temperature of $\sim$2.8 keV). A composite model consisting of two CIE plasma components plus emission lines from charge exchanges between cosmic-ray particles and interstellar hydrogens was proposed by Tanaka (2002) based on the "broadened" Fe $K\alpha$ emission lines detected by *ASCA*. Such a broadening was not detected in later observations (the present data and see Ebisawa et al. 2008), and thought to be an artificial effect probably caused by the instrument.

6.2.3 *Wide-Band Spectral Fitting*

As shown in Sect. 6.2.1, the hard X-ray spectrum of the GRXE is reproduced well with spectral models which have thermal nature (the IP model or a CIE plasma model). Besides, we reconfirmed after Kaneda et al. (1997), from the soft X-ray spectrum, the multi-temperature nature of the GRXE (Sect. 6.2.2). Based on these, in this section, we try to reconstruct the wide-band GRXE spectrum (Fig. 6.7) using as small number of physically plausible spectral components as possible.

6.2.3.1 Model Construction

As demonstrated in Sect. 6.2.2, the two collisional ionization equilibrium (CIE) components reproduced well the GRXE spectrum in the XIS energy band (2–10 keV). Therefore, this will be a nice start point when analyzing the wide-band GRXE spectrum over 2–50 keV, and hereafter, we designate this model as "Model 1" for simplicity. In addition to them, as noted above, we consider a contribution from the foreground diffuse soft X-ray emission. Based on recent reports (Yoshino et al. 2009; Masui et al. 2009; Kimura 2010), we assumed the surface brightness of this foreground emission to be 2.5×10^{-9} erg cm^{-2} s^{-1} sr^{-1}, and included it in the spectral model as a CIE plasma component with a temperature fixed at $kT = 0.7$ keV.

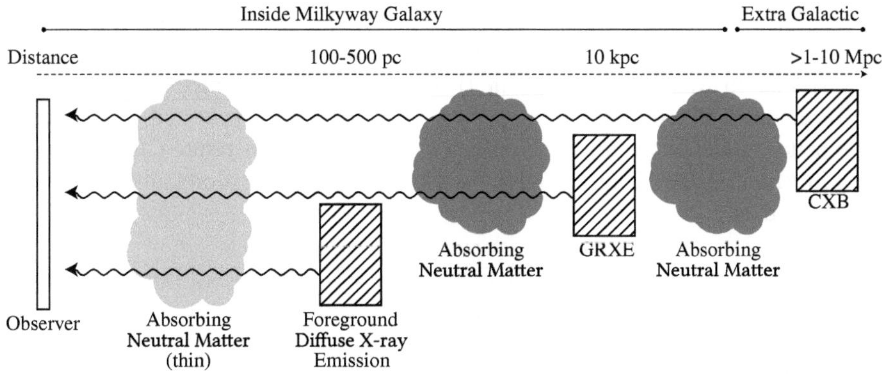

Fig. 6.15 Geometrical configuration of individual emission components (*hatched rectangles*) and absorbing interstellar (plus intergalactic) neutral matters (*gray* cloud-like objects). The *wavy lines* represent emitted X-rays which pass through the absorbing matters, and arrive to the observer

Practically, contribution from the foreground emission is only a few percent of that from the GRXE.

To visualize the location of individual emission and absorption components, Fig. 6.15 illustrates a geometrical configuration assumed in our analysis. Assuming this configuration, we summed up spectral models of the foreground diffuse emission, the GRXE, and the CXB, by multiplying photo absorption caused by the intervening absorbing matters. Since the two plasma components in the GRXE can come through different absorption columns densities, they were subjected to independent absorption factors. The abundances of the major elements (i.e., Si, S, Ar, Ca, and Fe) are allowed to vary individually, but are constrained to be the same between the two CIE components.

The Fe I Kα emission line could be a powerful tool for understanding the origin of the GRXE because some fraction of this line is surely produced via reprocessing (reflection) of hard X-rays by neutral Fe atoms in the atmosphere or accretion disks in WDs (Sect. 5.3.4). If Fe I Kα photons are emitted from these locations, its line shape is probably distorted by Compton scattering in the atmosphere or the gravitational redshift, and by analyzing the shape, we may be able to estimate fractional contributions from accreting WDs to the observed Fe I Kα line. However, this would require higher data quality (especially energy resolution better by an order of magnitude) and a detailed modeling which includes the reflection process. Therefore, in the present analysis, we approximate it by a Gaussian with its normalization left to vary freely.

The second model, Model 2, was constructed by replacing one of the two CIE plasma components of Model 1 with our IP spectral model (`IP_PSR`). This can be supported by the successful application of this model to the HXD/PIN GRXE spectrum (Sect. 6.2.1), and also by a suggestion that magnetic CVs are a major contributor to the GRXE (Revnivtsev et al. 2006, 2009). Non-magnetic CVs (dwarf novae), which are also thought to be major contributors to the GRXE below 10 keV, have spectral shape well explained by an optically-thin thermal emission from a hot plasma with a temperature distribution (e.g., Pandel et al. 2005). This condition is very similar to what assumed in our IP model, and as presented in Appendix C, the IP model well simulates their spectra. Therefore, we expect that Model 2 may have better modeling accuracy compared to the single-temperature CIE component used in Model 1 (see Sect. 6.3.1 for detailed discussion). Specifically, Model 2 is constructed as

$$\begin{aligned}\text{Model 2} = {} & \text{Foreground Diffuse Emission} + \text{CXB} + \text{Fe I K}\alpha\ \text{EmissionLine} \\ & + \text{CIE Plasma} + \text{IP}_{\text{P}}\text{SR}.\end{aligned}$$

The Fe abundance of the IP component was tied with that of the CIE Plasma component. Other abundance parameters were again allowed to vary individually (like Model 1). After the IP analysis results (Sect. 5.2.2), we applied a partial covering condition to the IP model.

We further subdivided this model into Model 2a and 2b, by treating the $M_{\rm WD}$ parameter of the IP component differently. In Model 2a, $M_{\rm WD}$ was fixed at 0.66 $M_{\odot}$ derived from the HXD/PIN spectral fitting (Sect. 6.2.1). The WD mass parameter of Model 2b was allowed to vary freely. If Model 2b gives a WD mass which is close to 0.66 $M_{\odot}$ the wide-band fitting can be considered to be fully consistent with that in the hard X-ray band.

6.2.3.2 Fit to the Wide-Band Spectrum

We first fit the models to the GRXE spectrum of Region 1 because it has higher counting statistics and covers wider energy range over 2–50 keV. As drawn in Fig. 6.16, all three models surprisingly well reproduced the wide-band spectra in this region. We list the best-fit parameters in Tables 6.5 and 6.6. Although the values of χ^2 are not yet small enough to make the fits formally acceptable, we consider again that these are caused by the uncertainty of the XIS response in the 2–4 keV band as explained in Sect. 6.2.2. If we include additional systematic errors (Sect. 6.2.2), the Model 2 fits became acceptable with null hypothesis probabilities larger than 1 %, while that of Model 1 was still unacceptable (probability of 0.2 %).

The fitting statistics of Model 1 and 2 (e.g., 2a) give a very low F statistic probability of $\sim 10^{-20}$, manifesting a better spectral reproducibility of Model 2 than Model 1. A similar (very low) probability is obtained from the fits with the mimicked systematic errors (using $\chi^2_{\nu,3\,\%}$ values in Tables 6.5 and 6.6). Actually, Model 2 better reproduces than Model 1 the XIS data at the S Kα line energies (2.5–2.7 keV), and the general slope of the HXD data. This is because CIE 2 of Model 1, with its temperature mainly determined by the HXD slope, under-predicts the Fe XXVI Kα line. As a result, CIE 1 of Model 1 is forced to be hotter so as to fill up the Fe-K line deficit. As a result, the model under-predicted the S XV Kα to S XVI Kα line

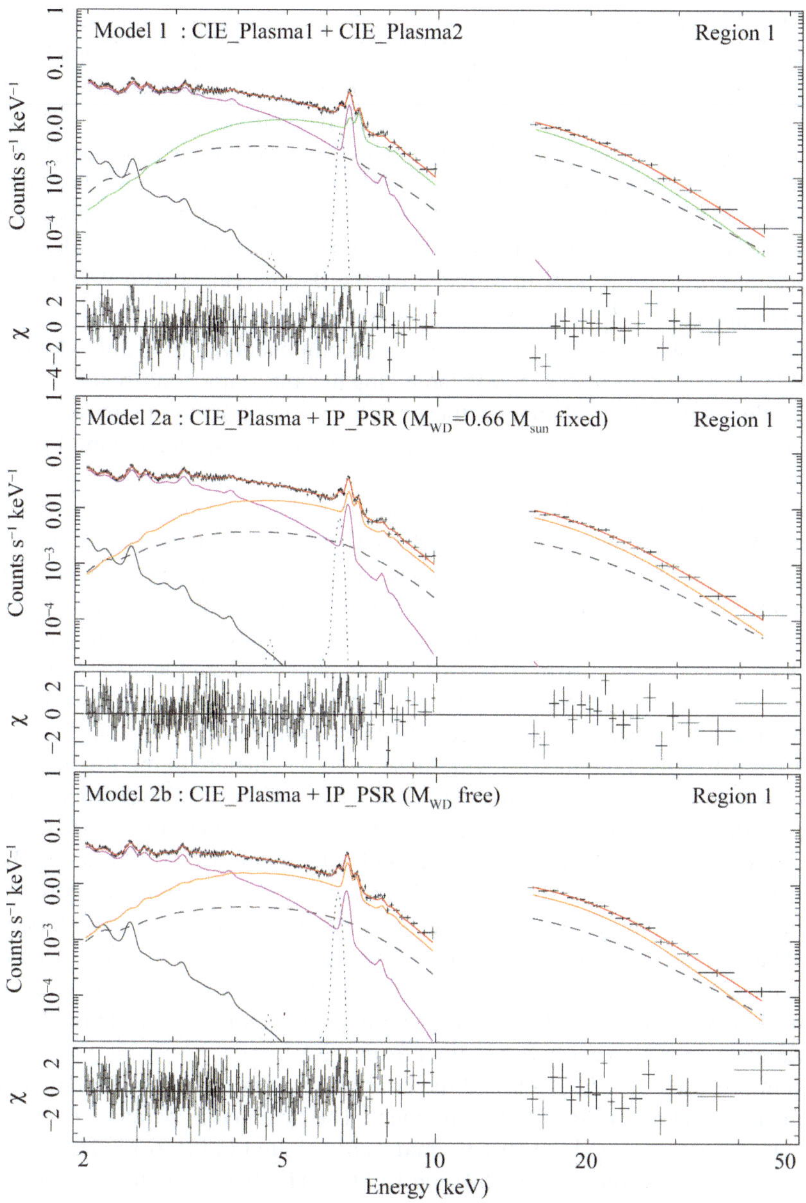

Fig. 6.16 The GRXE spectrum obtained from Region 1, fitted with Model 1 (*top panel*), Model 2a (*middle panel*), and Model 2b (*bottom panel*). *Black crosses* are the observed spectrum (NXB subtracted), and *red curves* are the sum of all model components. In the Model 1 fit, the two CIE plasma components are shown in *magenta* (lower temperature) and *green* (higher temperature). *Magenta* and *orange curves* the *lower two panels* are the CIE plasma and the IP model components, respectively. *Solid*, *dashed*, and *dotted gray curves* are components representing the foreground diffuse soft X-ray emission, the CXB, and Kα emission line from neutral Fe, respectively

ratio. We consider, based on these facts, that Model 2 is an appropriate model for the GRXE, within the present guiding principle of "as simple as possible" modeling.

The CIE 1 component of Model 1 and CIE of Model 2 yielded a plasma temperature of $kT = 1.4$–1.7 keV, and account for the soft continua plus Fe XXV Kα line with (almost) no Fe XXVI Kα photons. This is simply because such plasma temperatures are low to sustain the Fe XXVI ionization state (hydrogen-like), and only few Fe XXVI ions exist in the plasma. Most of the hard X-ray flux detected in the HXD/PIN is explained by the higher temperature components, CIE 2 with a temperature of $15.1^{+0.4}_{-0.7}$ keV in Model 1, or the IP component with WD masses of $0.66\ M_{\odot}$ and $0.48^{+0.05}_{-0.04}\ M_{\odot}$ in Model 2a and 2b, respectively.

Regardless of the models, the two GRXE components (CIE 1/2 in Model 1 or CIE/IP in Model 2) cross at 4–5 keV, indicating that the major contributor of the continuum flux of the GRXE changes at these energies. This is an important fact obtained from the wide-band spectroscopy without introducing any strong assumption on the emission model.

Based on the successful model fits in Region 1, we proceeded to fit the Region 2 spectrum. Since we consider that the origin of the GRXE and properties of individual components do not largely differ between Regions 1 and 2 (only a few degrees apart), we applied the same models to Region 2. Furthermore, to examine robustness of the model parameters obtained through the Region 1 fits, we fixed some parameters which mainly determines spectral shapes of the individual components; CIE plasma temperatures and $M_{\rm WD}$. Normalizations of the components were allowed to vary freely to take into account the surface brightness difference.

When we applied the models with these fixed parameters to the Region 2 spectrum, the overall spectral shape was moderately reproduced. However, all models left large residuals around emission lines at 2–3 keV. In fact, the fitting statistics were acceptable in none of the applied models with large reduced $\chi^2 > 2$ (the number of degrees of freedom of $\sim$350). Specifically, we observe positive and negative residuals in S XV Kα and S XVI Kα, respectively, which are due to mismatch between line fluxes of the data and those predicted by the CIE component with its temperature fixed at the Region 1 value; qualitatively, the Region 2 spectrum requires lower plasma temperature for the CIE component requiring more photons from the lower ionization state of S, namely S XV, than from S XVI.

To improve the fits, we loosened the constraints imposed on the model, and the plasma temperature of the CIE 1 component. The obtained fit parameters are listed in Tables 6.5 and 6.6, and the best-fit models are plotted in Fig. 6.17. The values of χ^2_ν are greatly improved to 1.28(752) and 1.65(354) in Regions 1 and 2, respectively, which are comparable to those of Region 1. As expected, lower plasma temperatures were obtained, and the positive and negative residuals decreased. The difference of these temperatures is discussed in the next section.

Figure 6.18 shows the best-fit Model 2b in Region 1 by removing the foreground emission, the CXB, and the interstellar absorption. A hump-like structure seen in the 10–30 keV band is due to the partially-covering dense absorption applied to the IP model.

Table 6.5 Result of the wide-band spectral fitting with Model 1*

	$n_{\rm H}$[a] (10^{22} cm^{-2})	$kT_{\rm CIE\ 1}$[b] (keV)	$n_{\rm H}$[c] (10^{22} cm^{-2})	$kT_{\rm CIE\ 2}$[b] (keV)	$Z_{\rm Fe}$[d] ($Z_\odot$)	χ^2_ν	$\chi^2_{\nu,\ 3\%}$[e]	$F_{\rm CIE\ 1}$[f] (2–10)	$F_{\rm CIE\ 2}$[f] (2–50)	R[g]
Region 1	$4.0^{+0.2}_{-0.2}$	$1.66^{+0.04}_{-0.04}$	$15.9^{+1.5}_{-1.3}$	$15.1^{+0.4}_{-0.7}$	$0.97^{+0.06}_{-0.06}$	1.45(754)	1.15	4.86	7.83	1.61
Region 2	$1.3^{+0.2}_{-0.2}$	$1.31^{+0.03}_{-0.03}$	–	–	$1.06^{+0.07}_{-0.08}$	1.85(353)	1.25	7.90	11.77	1.49

*Blank parameters are fixed at the Region 1 values (see text)
[a]Hydrogen column density of interstellar absorption
[b]Plasma temperatures of the CIE 1 and the CIE 2 components
[c]Additional absorption column density applied to the CIE 2 component
[d]Fe abundance of the CIE components
[e]Improved fitting statistics achieved when the XIS response uncertainty was virtually taken into account
[f]Model-predicted fluxes of the CIE 1 and 2 components in 10^{-9} erg cm^{-2} s^{-1} integrated over specified energy ranges and a $|l| < 10°$ and $|b| < 10°$ region
[g]Ratio of $F_{\rm CIE\ 2}$ to $F_{\rm CIE\ 1}$

Table 6.6 Result of the wide-band spectral fitting with Models 2a and 2b*

	n_H[a] (10^{22} cm^{-2})	kT (keV)	Z_{Fe}[b] ($Z_\odot$)	n_H[c] (10^{22} cm^{-2})	M_{WD}[d] ($M_\odot$)	$n_{H,\,PC}$[e] (10^{22} cm^{-2})	C_{PC}[f]	χ^2_ν	$\chi^2_{\nu,3\,\%}$[g]	F_{CIE}[h] (2–10)	F_{IP}[h] (2–50)	R[i]
Model 2a												
Region 1	$3.6^{+0.2}_{-0.3}$	$1.52^{+0.04}_{-0.04}$	$0.86^{+0.05}_{-0.04}$	$13.0^{+9.4}_{-1.3}$	0.66(fixed)	525^{+385}_{-259}	$0.22^{+0.09}_{-0.08}$	1.30(753)	1.04	3.87	8.60	2.22
Region 2	$1.2^{+0.2}_{-0.2}$	$1.21^{+0.04}_{-0.03}$	$0.82^{+0.05}_{-0.05}$	–	–	–	–	1.62(354)	1.11	7.18	12.65	1.76
Model 2b												
Region 1	$3.2^{+0.4}_{-0.6}$	$1.44^{+0.06}_{-0.07}$	$0.73^{+0.06}_{-0.05}$	$11.8^{+8.6}_{-1.8}$	$0.48^{+0.05}_{-0.04}$	490^{+109}_{-103}	$0.62^{+0.09}_{-0.11}$	1.28(752)	1.02	3.23	8.59	2.66
Region 2	$1.1^{+0.2}_{-0.2}$	$1.17^{+0.04}_{-0.04}$	$0.67^{+0.04}_{-0.04}$	–	–	–	–	1.65(354)	1.13	6.49	12.48	1.92

*Blank parameters are fixed at the Region 1 values (see text)
[a] Hydrogen column density of interstellar absorption
[b] Fe abundance
[c] Additional absorption column density applied to the IP component
[d] WD mass of the IP model
[e] Hydrogen column density of partial-covering dense absorption in the IP model
[f] Partial-covering fraction
[g] Improved fitting statistics achieved when the XIS response uncertainty was virtually taken into account
[h] Model-predicted fluxes of the CIE and IP components in 10^{-9} erg cm^{-2}s^{-1} integrated over specified energy ranges and a $|l| < 10°$ and $|b| < 10°$ region
[i] Ratio of F_{IP} to F_{CIE}

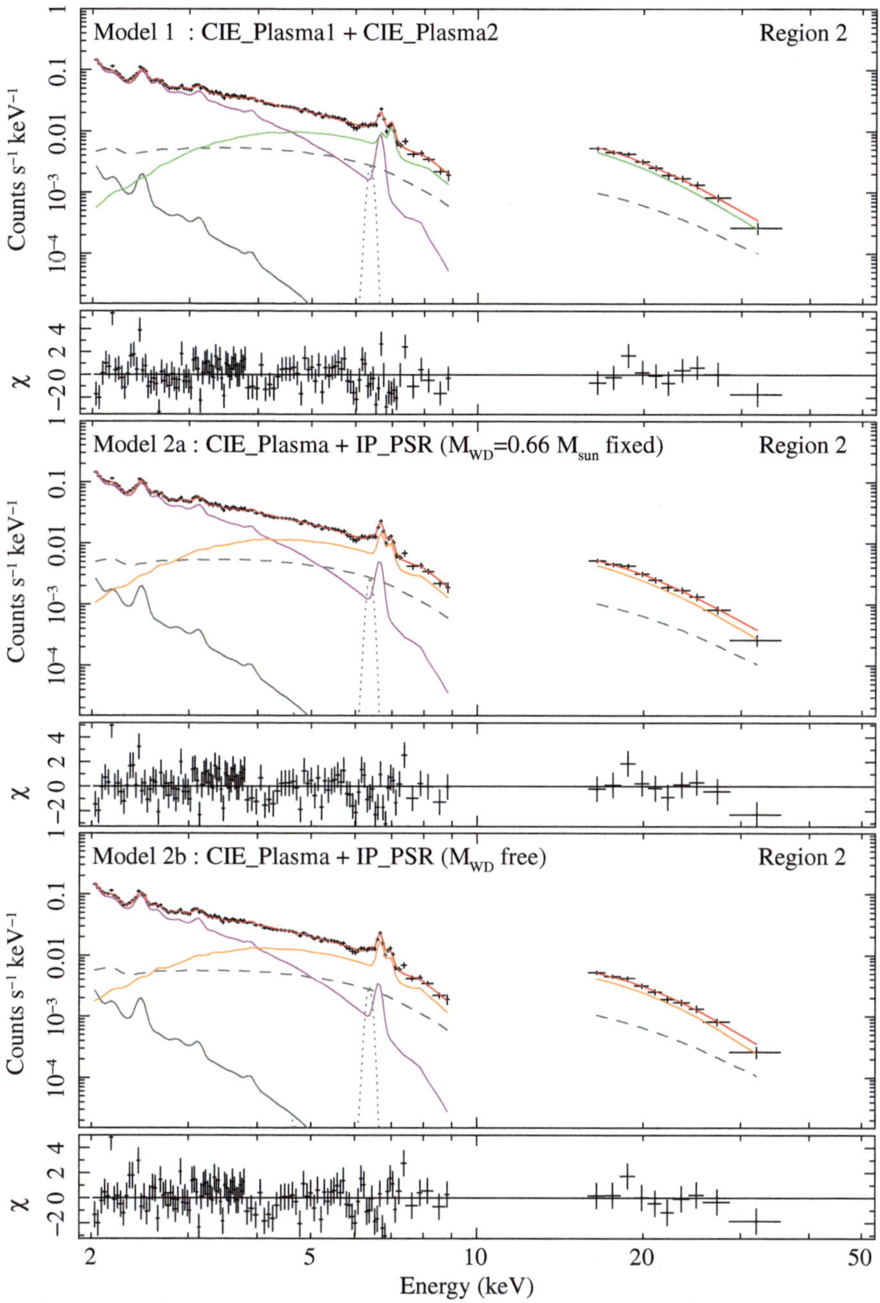

Fig. 6.17 The same as Fig. 6.16, but the spectra of Region 2 are shown overlaid with the best-fit model in which the plasma temperature of the lower-temperature CIE component was allowed to freely vary

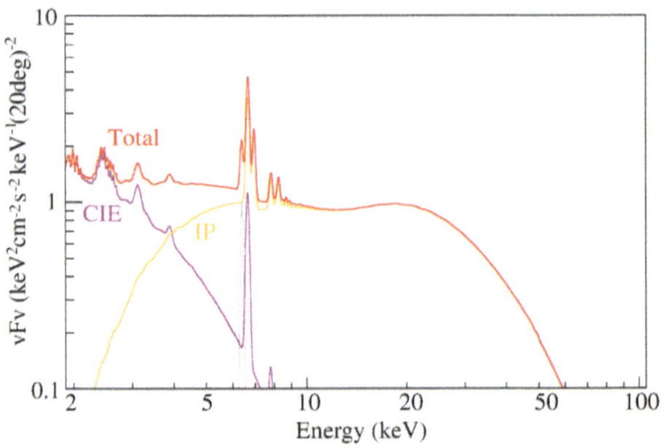

Fig. 6.18 Same GRXE model spectrum as the best-fit Model 2b in Fig. 6.16, but shown after removing the detector response and the interstellar absorption. The vertical axis corresponds to the spectral energy density or νF_ν, the units often used in X-ray studies. Emission region is assumed to the central 20° × 20° region around the Galactic center. For better visibility of emission lines, the spectrum is smoothed with a Gaussian function which has a full width at half maximum of 150 eV. The CXB and foreground components are omitted

In qualitatively summarizing the fit results, we emphasize a fact that these fits clearly showed, for the first time, that the GRXE spectrum up to 50 keV can be described well by thermal models; CIE 2 or the IP model in Models 1 and 2. This is a strong support for the stellar origin of the GRXE because, as discussed below, derived plasma temperatures (or WD mass) are consistent with values expected from this scenario, and a plasma with a temperature higher than 10 keV faces severe confinement problem in the "Diffuse" scenario. Combining the HXD/PIN GRXE fit (Sect. 6.2.1) and the wide-band spectral modeling shown here, we conclude that our data does not require to introduce a non-thermal component (e.g., power law) in the fits.

6.2.3.3 Abundance Profile

We also fitted the wide-band spectra extracted from individual observations (Sect. 6.1.4) using the constructed models without fixing the model parameters. Since counting statistics are limited, we represented abundances of S, Si, Ar, Ca, and Fe, by a single abundance parameter Z assuming the solar composition. Due to the limited spectral quality, fits were acceptable in all cases, and yielded generally the same result as the case of the summed wide-band spectrum.

As shown in Fig. 6.19, the metal abundance was then found to decrease as away from the Galactic plane. Similar tendency has been reported by Kaneda et al. (1997) using GRXE mapping observations performed over $(l, b) = (28.5°, 0°)$~$(28.5°, 1.7°)$. Note that the abundance value in the figure is systematically smaller than the iron abundance $Z_{\rm Fe}$ listed in Tables 6.5 and 6.6. This is probably an artificial

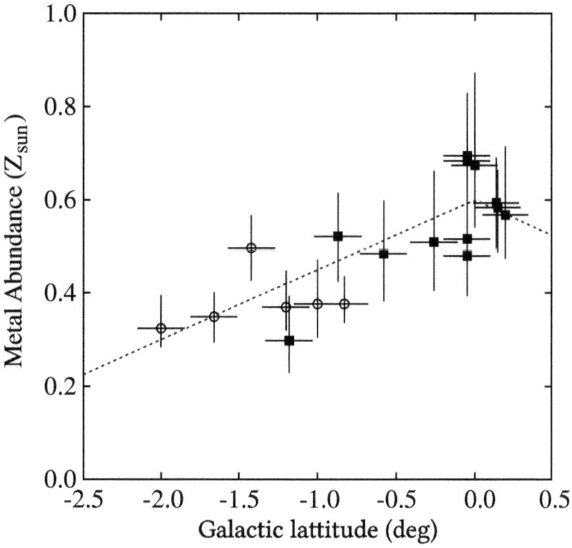

Fig. 6.19 Metal abundance profile along Galactic latitude. The filled rectangles and open circles correspond to values derived from individual observations in Region 1 and Region 2. The *vertical bars* represent statistical fitting errors at 90 % confidence level, whereas the *horizontal error bars* represent the XIS field of view ($18' \times 18'$). The *dashed line* is a tentative function $y = 0.15 \times |x| + 0.6$

effect caused by the simplification (the single abundance parameter). Therefore, absolute values of individual data points are of minor physical importance, whereas their relative ratios may reflect an actual spatial distribution of the metal abundance.

6.2.3.4 Surface Brightness of the GRXE

In Tables 6.5 and 6.6, we presented fluxes of individual spectral components predicted by the best-fit models. The value is normalized to a rectangular sky region, $|l| < 10°$ and $|b| < 10°$, i.e., $20° \times 20°$ around the Galactic center. The energy ranges for the fluxes were adjusted to typical ones often used in X-ray studies.

Besides these integrated fluxes, we also calculated local GRXE surface brightness (i.e., flux density per unit solid angle) at certain sky regions in Regions 1 and 2. Two fields of view of Obs. IDs 503076010 and 504089010 (see Fig. 6.1) are selected as representatives of the on-plane and the off-plane observations, respectively. To derive intrinsic surface brightness of the GRXE, the foreground and the CXB components were removed in the calculation, and the absorption component was also eliminated to correct the interstellar absorption which especially affects observed flux below 3–4 keV. Table 6.7 lists resulting surface brightness values.

The values in the on-plane region (Obs. ID 503076010) are larger than that of the off-plane region (Obs. ID 504089010) by a factor of 1.5. This implies that the GRXE is intrinsically brighter in on-plane regions, as was expected from the assumed surface brightness model (Fig. 6.9). These values are almost consistent with the surface brightness by Revnivtsev et al. (2006) using *RXTE*/PCA around the Galactic center ($|l| < 4°$), and larger by 5–6 times than that reported by Kaneda et al. (1997) at the off-center Galactic disk region $(l, b) = (28.5°, 0°)$. Since we are assuming that the GRXE surface brightness is proportional to the stellar density integrated along the

Table 6.7 Surface brightness of the GRXE calculated from the Model 2b fits in the on-plane and the off-plane regions

Obs. ID	Coordinate[a]	Surface brightness[b]	
	(l, b)	(2–10 keV)	(2–50 keV)
503076010	$(-1.50°, +0.15°)$	1.50	2.44
504089010	$(-0.05°, -1.20°)$	1.08	1.51

[a] Aim point of the observation in the Galactic coordinate
[b] Absorption-corrected surface brightness of the GRXE integrated over the specified energy ranges, shown in units of 10^{-10} erg cm^{-2} s^{-1} deg^{-2}. Statistical fitting errors of $\sim$20 % associate with these values according to the fit result (Table 6.6)

line of sight, this difference can be simply understood as higher stellar density in the Galactic bulge region (aim points of our GRXE observations) than in the off-center region (Dwek et al. 1995).

6.2.4 Systematic Uncertainties Involved in the Present Analysis

In this section, we review several systematic uncertainties which our analysis potentially involves. As explained below, however, none of them can modify the essential part of the result that the wide-band GRXE spectrum can be explained by the two thermal components showing beautiful agreements on physical parameters, such as the plasma temperature, with its proposed origins, namely active binaries and accreting WDs.

6.2.4.1 Surface Brightness Modeling

When calculating the instrumental responses (Sect. 6.1.6), we assumed the same surface brightness model (G3 model) as used in Revnivtsev et al. (2006) and Krivonos et al. (2007). In their papers, the model nicely reproduced observed surface brightness of the GRXE globally over $|l| < 30°$ and $|b| < 15°$. Our GRXE count rates also nicely correlate with those predicted by this model (Fig. 6.10). However, there might be some small scale fluctuation of the surface brightness which could affect our instrumental response calculation. Therefore, to assess this possible uncertainty, we performed spectral fitting again using instrumental responses calculated based on a different surface brightness model.

We utilized a model consists of two dimensional Lorentzian with full widths at half maximum of 21° and 1.2° which were used in Türler et al. (2010) to model the GRXE surface brightness. Obtained best-fit parameters were consistent with the values reported above within statistical fitting errors. Only difference was a normalization parameter of each spectral component, and they are consistent with relative scaling between the two surface brightness models. Based on this, we can conclude that our result does not greatly depend on assumed surface brightness models as long as we use realistic ones.

6.2.4.2 Surface Brightness of the Foreground Diffuse X-ray Emission

When constructing the wide-band GRXE spectral model, we took into account the foreground diffuse soft X-ray emission (Kuntz and Snowden 2008; Yoshino et al. 2009; Masui et al. 2009) using a surface brightness of 2.5×10^{-9} erg cm^{-2} s^{-1} sr^{-1} (2–10 keV). Because this is a prediction based on observations of distant sky regions, the value could have an error as large as a factor of several according to the comparison of two sky regions $(l, b) = (123.0°, 0.0°)$ and $(235.0°, 0.0°)$ by Kimura (2010).

Although flux from this component is only a few percent of that of the detected GRXE, the lower plasma temperature derived from the wide-band fitting might be affected by this. Therefore, we examined how the result changes depending on the assumed surface brightness of the foreground emission. Figure 6.20 shows two examples with decreased and increased foreground emission components (50 and 200 % of the original surface brightness). Model 2b fitted the data at almost the same level as did before, and the flux difference caused by the scaled foreground emission is compensated mainly by slightly changing the normalization of the lower-temperature CIE component and the interstellar absorption column density. The best-fit plasma temperature and WD mass were quite consistent with the original result within errors; $1.42^{+0.06}_{-0.08}$ keV and $0.46^{+0.04}_{-0.04}$ M_{WD} in the 50 % case, and $1.48^{+0.06}_{-0.06}$ keV and $0.49^{+0.06}_{-0.05}$ M_{WD} in the 200 % case. Thus, although we do not know the surface brightness of the foreground diffuse emission precisely in the present sky region, our result has little dependence on this uncertainty.

6.2.4.3 Confusion of "Resolvable" Point Sources

It is clear that our GRXE spectrum contains signals from faint point sources which can be resolved using instruments with higher spatial resolution, for example the *Chandra*/ACIS or the *XMM-Newton*/EPIC CCDs. However, the purpose of the present analysis is not to identify each point source, but to spectroscopically decompose constituents of the GRXE. This cannot be done in individual spectra of resolved point sources because of very low counting statistics (a few to several tens of photons per one point source in an ultra-deep *Chandra* observation of ~1 Ms for example).

As described as one of the target selection criteria (Sect. 6.1.1), we defined the GRXE as X-ray signals integrated inside a certain field of view that does not contain an apparent point source brighter than 0.1 mCrab ($\sim 10^{12}$ erg cm^{-2} s^{-1} in the 2–8 keV band, or an intrinsic X-ray luminosity of 8×10^{33} erg s^{-1} when located at the Galactic center). The detected GRXE flux is ~10 times higher than this limiting flux. From this definition, the above mentioned *contribution* from faint point sources which could be resolvable with the other instruments is not *noise* but *GRXE signals* in our analysis.

Besides, as far as we know, the limiting flux (0.1 mCrab) to a contaminating *bright* point source is the lowest or the tightest among those of recent GRXE studies

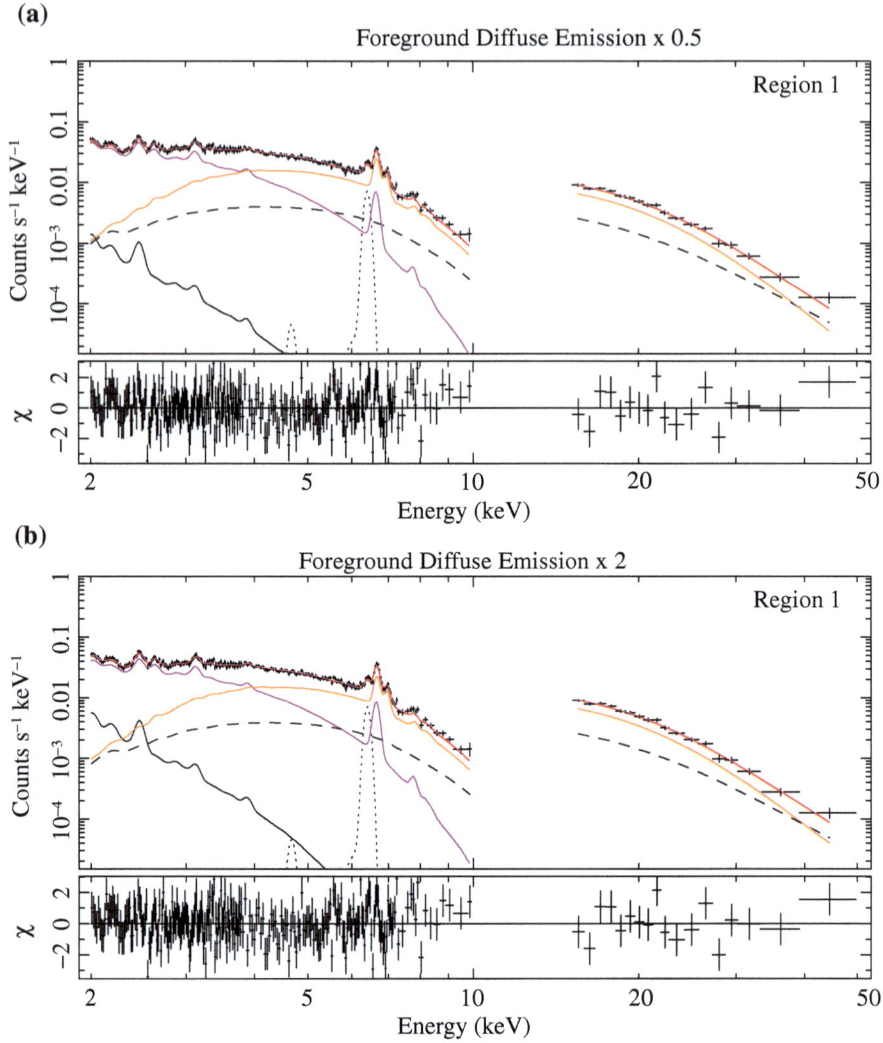

Fig. 6.20 The same GRXE spectrum of Region 1 as shown in Fig. 6.16 fitted with Model 2b, but with the foreground diffuse emission being scaled by factors of (**a**) 0.5 and (**b**) 2

in the hard X-ray band ($E > 10$ keV; Revnivtsev et al. 2006; Türler et al. 2010). In addition to the inspection of the XIS images, we also checked the light curves of the count rates of individual observations, and found no significant time variability. This ensures that no bright transient activity, which could affect the observed GRXE spectrum, occurred during the observations.

6.3 Discussion

6.3.1 Interpreting the Spectral Parameters

Since the "Point Source" scenario of the GRXE origin is our working hypothesis (Sect. 2.1.6), in this section, we interpret the model parameters obtained above (Tables 6.5 and 6.6) by comparing them with typical values expected from individual contributors assumed in this scenario.

6.3.1.1 Plasma Temperature and White Dwarf Mass

The lower-temperature CIE component (magenta in Figs. 6.16 and 6.17) resulted in ranges of $kT = 1.4$–1.7 keV and 1.2–1.3 keV in Region 1 and 2, respectively. This temperature is beautifully consistent with those seen in X-ray spectra of active binaries (e.g., Favata et al. 1997; Covino et al. 2000) which are binary stars harboring two late-type (or "normal") stars like the Sun, and exhibit X-ray emission via magnetic activities in their coronae. Small temperature difference between the two regions ($\sim$0.3–0.4 keV) can be naturally understood if there is unresolved contribution from young supernova remnants (SNRs) which tend to exhibit thermal X-ray spectra of plasma (electron) temperatures of $kT \sim 2$–4 keV (e.g., Kinugasa and Tsunemi 1999 for *Kepler*'s SNR; Tamagawa et al. 2009 for *Tycho*'s SNR). Recent discoveries or possible discoveries of new SNRs in the Galactic center region (Sawada et al. 2009; Nobukawa et al. 2008), especially in on-plane regions such as Region 1, support this speculation on the contribution from unresolved SNRs.

The temperature of the CIE 2 component of Model 1 ($15.1^{+0.4}_{-0.7}$ keV), which was constrained well thanks to the hard X-ray continuum shape measured by HXD/PIN, is consistent with a representative plasma temperature (10–20 keV) of magnetic CVs. The Model 2a fit successfully reproduced the wide-band spectrum although a WD mass parameter was fixed at the value derived from the HXD/PIN analysis ($0.66^{+0.09}_{-0.07}$ $M_{\odot}$; Sect. 6.2.1). When it was unfixed in Model 2b, a similar but slightly lower WD mass of $0.48^{+0.05}_{-0.04}$ $M_{\odot}$ was obtained. This reduction in WD mass is also explained by an unresolved contributor from other types of CVs, especially dwarf novae, whose spectra can also be reproduced with a CIE plasma model with lower temperatures ($kT \sim$ a few–10 keV; see Appendix B) than those of magnetic CVs.

Is the reduction of the WD mass parameter by $\sim$0.10–0.15 $M_{\odot}$ realistically possible with the contribution of dwarf novae? To test this, we simulated a composite wide-band GRXE spectrum (excluding the low-temperature CIE component) by adding spectral models of dwarf novae and IPs. An average plasma temperature of $kT = 5$ keV and an average WD mass of 0.6 $M_{\odot}$ were assumed for the dwarf nova and the magnetic CV components, respectively (see Appendix C for the temperature of dwarf novae). Relative fluxes of individual components were adjusted to (dwarf novae):(magnetic CV) = 0.3:1 according to X-ray emissivities of these

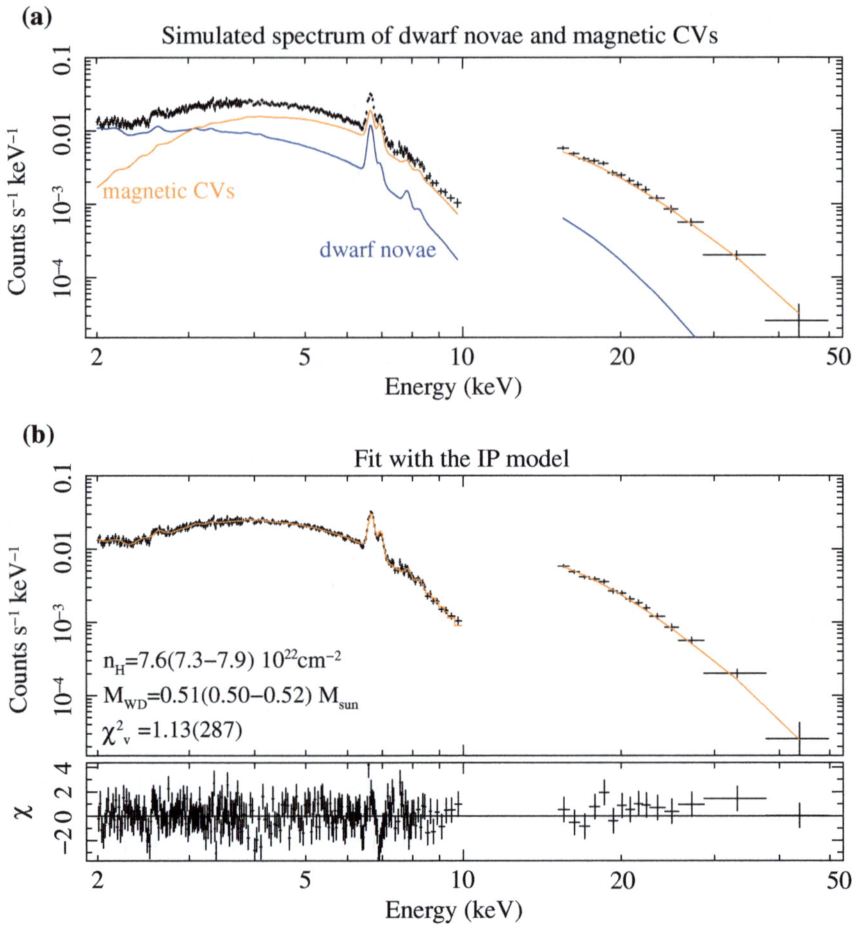

Fig. 6.21 **a** A simulated GRXE spectrum (*black crosses*) composed of spectral models of the dwarf nova (*blue curve*) and the magnetic CV (*orange curve*; the IP model). **b** The same spectrum as panel **(a)**, but fitted with the IP model (*orange curve*)

categories measured by Sazonov et al. (2006). For better simulating the observed spectrum, we also applied photo absorption models with hydrogen column densities of $n_{\rm H} = 3 \times 10^{22}$ cm^{-2} (Galactic interstellar value for dwarf novae) and 10×10^{22} cm^{-2} (dense intrinsic absorption for magnetic CVs). Simulated count rates were adjusted so as to match those of the observed data. Figure 6.21a presents a thus produced spectrum. This spectrum mimics the GRXE spectrum (subtracting the low-temperature CIE component).

We performed a similar fit to this spectrum using only the IP model, and obtained an acceptable fit which yields $\chi^2_\nu = 1.13(287)$. The best-fit model is plotted in Fig. 6.21b, and it gave the best-fitting WD mass of $0.51 \pm 0.01\ M_\odot$ and the absorption column density of $n_{\mathrm{H}} = 7.6 \pm 0.3 \times 10^{22}$ cm^{-2}. Thus, even when the dwarf nova component is missing in the fit model, the composite simulated spectrum can be well reproduced with the IP model with the WD mass parameter, which is slightly reduced from the assumed value of 0.6 $M_{\odot}$ ($\Delta M_{\mathrm{WD}} = 0.09$ $M_{\odot}$). This supports the above discussion on the reduced WD mass obtained in the wide-band spectral analysis.

6.3.1.2 Absorption Column Density

An absorption column density of $n_{\mathrm{H}} \sim 3\text{–}4 \times 10^{22}$ cm^{-2} was obtained in Region 1, while that of Region 2 was $\sim$1.1–1.3 $\times$ 10^{22} cm^{-2}. This is consistent with general understanding of the interstellar gas distribution in the Galaxy (e.g., Kalberla et al. 2005); i.e., more in the Galactic plane, less above and beneath the plane.

The hotter CIE component (Model 1) and the IP component (Model 2) required stronger absorption ($n_{\mathrm{H}} = 11\text{–}16 \times 10^{22}$ cm^{-2}). In the IP study (Table 5.2), we also obtained the similar result. Absorption caused by the pre-shocked accreting matter itself is widely accepted as the cause of this absorption, and these column densities exceeding 1×10^{23} cm^{-2} are achievable within realistic accreting geometry; for example, pre-shock gas with a typical density of $\sim 5 \times 10^{-9}$ g cm^{-3} (equivalent to 2.3×10^{15} H cm^{-3}; see Fig. 3.5) with a line-of-sight length of 500 km (one$\sim$several times of the shock height) easily amounts $\sim 1 \times 10^{23}$ cm^{-2}.

Note that in the partial covering absorption, the hydrogen column density and the covering fraction are coupled each other, and their individual values are rather ambiguous compared to a plasma temperature or a WD mass which clearly determines a spectral shape and emission line intensities. Therefore, although the covering fractions of Models 2a and 2b differ each other, we consider that they are not directly reflecting actual difference of the covering fraction in magnetic CVs.

6.3.1.3 Fe Abundance

Fe abundances of $\sim$1(0.9–1.1) $Z_{\odot}$ and $\sim$0.8(0.7–0.9) $Z_{\odot}$ were obtained in the Model 1 and the Model 2 fits, respectively. If the GRXE has a stellar origin, i.e., a collection of numerous faint X-ray point sources, these values represent typical Fe abundance of individual sources. If there is a mass donor in the source (i.e., if it is an X-ray emitting binary), the abundance corresponds to that of the companion star as was the case in IPs (Sect. 5.4.3).

When compared to the Fe abundance measured in the nearby IPs (0.33 ± 0.14 $Z_{\odot}$; Sect. 5.4.3), these values are larger by at least a factor of two. Considering that the present GRXE data is constructed from the observations performed within a few degrees around the Galactic center, this enhancement seems consistent with the general idea of metal-abundance gradient in the Galaxy (higher toward the Galactic center; Andrievsky et al. 2002b for review). According to recent measurement by Andrievsky et al. (2002a), Fe abundance (of Cepheid-type variable stars) seems to be higher by $\sim$2–3 times in central regions of our Galaxy (galactocentric distance $< \sim 3$ kpc) compared to that of the solar vicinity, and this is roughly consistent with the enhancement we obtained.

6.3.2 Comparison of Observed GRXE Spectral Shapes

Figure 6.22a compares spectral shapes of the present study alongside those reported by previous studies in hard X-rays. Because studied sky regions are different among the reports, we scaled the data points so that the *RXTE* spectrum (3–20 keV) match our model. Note that these studies integrated the GRXE from rather wide sky regions $|l| < 30°$ and $|b| < 15°$. Although the spectra agree well below 10 keV, hard X-ray spectra above 10–20 keV exhibit larger deviations. Note that Krivonos et al. (2007) and Türler et al. (2010) reported different spectral shapes using the same instrument, but from data of different observations. According to a reanalysis of the *INTEGRAL*/IBIS data using an updated method, it was recently suggested by Krivonos (2010; private communication) that the previous GRXE flux, shown in Fig. 6.22a, should be increased at least by 10–20 %. Figure 6.22b shows the same GRXE spectrum measured with *INTEGRAL* but shifting the normalization by a factor of 1.2. The deviation, of course, becomes smaller, and two spectra (ours and that of rescaled from Krivonos et al. (2007) almost agree in normalizations and spectral shapes.

Based on this, we again confirm that the present GRXE spectrum measured with *Suzaku* is almost free from unidentified contamination from bright point sources, and *actually contains the GRXE signals*. The still remaining ambiguity of the hard X-ray GRXE spectral shape and normalization, especially that observed between the present result and the rescaled spectrum of Krivonos et al. (2007) at a $\sim$10 % level, will be eliminated by coming hard X-ray imaging instruments with collecting mirrors, such as the *NuSTAR* satellite (Harrison et al. 2010) or the Hard X-ray Imager onboard the *ASTRO-H* satellite (Takahashi et al. 2008), which cover up to 70 keV with sensitivities higher by an order of magnitude or more than those achieved in the currently available instruments.

6.3.3 The Number Density of the Unresolved Hard X-ray Point Sources

6.3.3.1 The X-ray Luminosity Function

As presented in Table 6.7, we measured the GRXE surface brightness in 2–50 keV band. Especially, the value in the hard X-ray band (>10 keV) can be used to constrain the population of unresolved point sources, mostly IPs (see Fig. 2.4). Using the

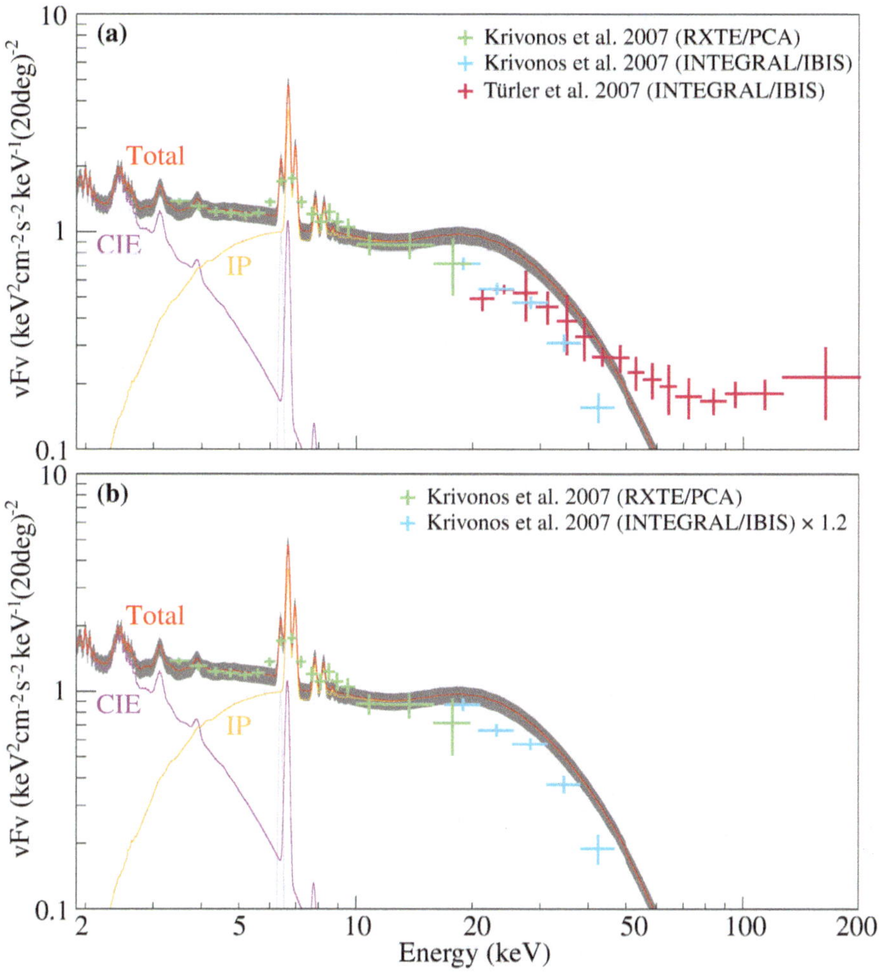

Fig. 6.22 **a** The same GRXE model spectrum as Fig. 6.18, but data points taken from previous studies are overlaid; *green* and *blue crosses* from Krivonos et al. (2007) and *magenta crosses* from Türler et al. (2010). The *gray strip* indicates ±10 % of our model which roughly corresponds to statistical fitting errors of our analysis. **b** The same as panel **(a)**, but *INTEGRAL* spectrum (*blue crosses*) are shifted upward by 20 % according to a recent report by Krivonos (2010; private communication)

GRXE flux observed by HXD/PIN, we derive their population, or the so-called X-ray luminosity function (i.e., number density versus intrinsic luminosity). In the present calculation, we assume that all the point sources are located at a distance of 8 kpc from the Sun (i.e., at the Galactic center region) as done in many previous studies (e.g., Muno et al. 2009).

Generally, a luminosity function of X-ray stars N can be expressed as a power-law function, and therefore we write N using the intrinsic luminosity L

$$N(>L) = N_0 \left(\frac{L}{L_0}\right)^{-\alpha}. \tag{6.4}$$

Here, $N(>L)$ means the surface number density of (unresolved) point sources which have an intrinsic luminosity L, and has a dimension of [(number) (solid angle)$^{-2}$]. N_0 and L_0 are constant factors for scaling, and the latter can be arbitrarily fixed. The parameter α denotes the luminosity dependence of N.

Now, we can calculate an X-ray surface brightness S predicted from the source population N by integrating a product of the number of sources (i.e., derivative of the cumulative density N) and the luminosity L as

$$S = \int_{L_{\max}}^{L_{\min}} \frac{\mathrm{d}N(>L)}{\mathrm{d}L} L \,\mathrm{d}L. \tag{6.5}$$

The integration is performed over a luminosity range $L_{\min} - L_{\max}$ where $\mathrm{d}N(>L)/\mathrm{d}L$ is non-zero.

For quantifying the luminosity function from the present GRXE flux, we determined $L_{\min}$ and $L_{\max}$ as follows. The present GRXE study does not contain bright known X-ray sources which have energy fluxes higher than $\sim 10^{12}$ erg cm^{-2} s^{-1} in the 2–8 keV band, or equivalently ~0.1 mCrab. The flux limit corresponds to 1.7×10^{12} erg cm^{-2} s^{-1} when expressed in the 12–50 keV band (i.e., HXD/PIN energy coverage). If we consider a point source located at 8 kpc from the Sun (i.e., at the Galactic center region), the flux means an intrinsic luminosity of 1.3×10^{34} erg s^{-1} in the same energy range. This should be regarded as $L_{\max}$. Recent deep *Chandra* observations revealed that there exist X-ray sources with luminosities as low as 10^{30} erg s^{-1} in the 0.5–7 keV band (Revnivtsev et al. 2009). Another important fact is that Muno et al. (2009) securely measured the shape of the luminosity function down to 3×10^{31} erg s^{-1} in the 2–10 keV band. This value provides stringent "upper limit" for $L_{\min}$; i.e., its actual value should be much lower probably one order of magnitude because there is no sign of break, or turn off, in their luminosity function. Since the actual value of $L_{\min}$ is still unclear, we use these actually measured values as $L_{\min}$ in the integration of Eq. (6.5).

Before the integration, these luminosity values should be converted to those in the 12–50 keV band which we concentrate on in the present calculation. Since most of the unresolved point sources are thought to be accreting WDs (especially magnetic ones), they have hard spectra (i.e., high plasma temperatures; Muno et al. 2004). We assumed, for the unresolved sources, a typical X-ray spectrum consisting of a single-temperature CIE plasma model with $kT = 20$ keV suffered from the interstellar absorption of $n_{\mathrm{H}} = 6 \times 10^{22}$ cm^2 (typical value for the Galactic center region). Based on this spectral shape, luminosities measured in the 0.5–7 and 2–10 keV

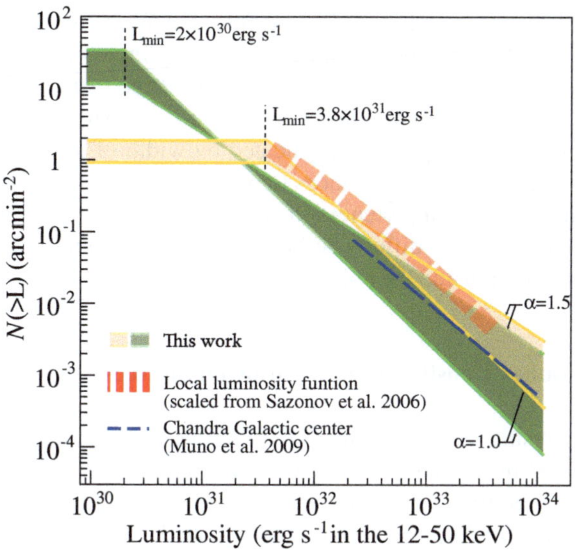

Fig. 6.23 X-ray luminosity functions of unresolved point sources, which make up the GRXE in the higher energy band above 10 keV, calculated from the HXD/PIN GRXE flux assuming $L_{\min} = 2 \times 10^{30}$ (*green* region) and 3.8×10^{31} erg s^{-1} (*orange* region) in the 12–50 keV band. The *two solid lines* which enclose each region have differently-assumed power-law indices of $\alpha = 1.0$ and 1.5. The *blue dashed line* represents a luminosity function that was actually measured by *Chandra* in similar sky region ($2° \times 0.8°$ around the Galactic center) in the soft X-ray energy band 2–10 keV (Muno et al. 2009). *Red dashed curve* is a local X-ray luminosity function of CVs (Sazonov et al. 2006), scaled using the stellar mass contained in the HXD/PIN field of view ($M \sim 8.8 \times 10^7\ M_\odot$). The line width shows estimated uncertainty involved in the conversion factor and the luminosity function itself

band can be converted to a value in the 12–50 keV by multiplying 2.03 and 1.27, respectively. Therefore, assumed $L_{\min}$ values become 2×10^{30} and 3.8×10^{31} erg s^{-1}.

The present HXD/PIN measurement gives the GRXE surface brightness, for example, $S = 2.34 \times 10^{35}$ erg s^{-1} (PIN FOV)$^{-1}$ [2] in Obs. ID 504001010 centered at $(l, b) = (-1.47°, -0.26°)$. Since S, or $N(> L)$ inside it, includes two unknown parameters α and N_0, an additional constraint on one of the two parameters should be placed to determine $N(> L)$. Previous studies (Ebisawa et al. 2005; Muno et al. 2009) have revealed that the luminosity function of faint Galactic X-ray point sources has $\alpha = 1.0 - 1.5$. Based on this, we calculate $N(> L)$ using two representative indices $\alpha = 1.0$ and 1.5.

Figure 6.23 shows the calculated luminosity functions of the unresolved sources. The two filled regions present two representative cases with lower luminosity limits of $L_{\min} = 2 \times 10^{30}$ erg s^{-1} (green region) and 3.8×10^{31} erg s^{-1} (orange). The luminosity functions are consistent with what was actually measured in similar sky regions by *Chandra* (Muno et al. 2009). In Fig. 6.23, we also compared our result with

[2] The effective field of view of HXD/PIN ($34' \times 34'$). See Sect. 4.1.2.

a local X-ray luminosity function of CVs measured in the solar vicinity ($\lesssim$500 pc) in the 3–20 keV band (Sazonov et al. 2006). When adopting their result, we took data points from their Fig. 13 (see Fig. 2.3 in the present thesis), and multiplied them by $8.8 \times 10^7\ M_\odot / 7 \times 10^8\ M_\odot$ to calculate source number density in the HXD/PIN field of view (see below). Although error (indicated by width of the red curve) is rather large due to uncertainties of the distances to observed sources, stellar mass enclosed in the local volume, and completeness, it roughly agrees with our allowed regions indicating that there is little difference between the X-ray luminosity functions of the Galactic bulge region and the solar vicinity.

6.3.3.2 The Number Density of the Unresolved Sources

From the above result, we can estimate the number density of unresolved sources which should exist in the field of view of HXD/PIN. The saturating values of the luminosity functions give the surface number densities of 11.6–34.2 arcmin^2 and 0.92–1.88 arcmin^{-2} for $L_{\mathrm{min}} = 2 \times 10^{30}$ erg s^{-1} and 3.8×10^{31} erg s^{-1}, respectively. By multiplying its effective sky coverage of $34' \times 34' = 1156$ arcmin^{-2}, we obtain 1.3–3.9 $\times 10^4$ and 1.1–2.2 $\times 10^3$ sources inside the HXD/PIN field of view depending on the assumed lower limit luminosities and the power-law indices. As explained above, we consider that the case with $L_{\mathrm{min}} = 3.8 \times 10^{31}$ erg s^{-1} only give "lower limit" of the source density. Therefore, at least $\sim 10^3$, and probably $\sim$a few $\times\ 10^4$ sources composing the GRXE detected by HXD/PIN.

Is not this source density too high for the Galactic source population? The answer is simply *no*. These values are rather consistent with one predicted from the deep *Chandra* observation (Revnivtsev et al. 2009); based on their source density $\sim$20 arcmin^{-2} (see Fig. 2.6a), 2.3×10^4 sources are expected in the HXD/PIN field of view.

The nice correlation of the GRXE surface brightness and that of the near infrared diffuse emission (Sect. 2.1.1) provides independent check for the derived source density. Based on the GRXE study in large sky area, Revnivtsev et al. (2006) estimated a ratio of the GRXE luminosity and the stellar mass (traced by the near infrared emission) as

$$S_{\mathrm{X}}/M \sim (3.5 \pm 0.5) \times 10^{27}\ \mathrm{erg\ s^{-1}}\ M_\odot^{-1}, \tag{6.6}$$

where S_{X} is the GRXE surface brightness measured in the 3–20 keV band, and M is the total stellar mass integrated in a volume which emits the GRXE. Multiplying a factor of 0.76 to the right hand side of Eq. (6.6) yields a corresponding value in the 12–50 keV following the same conversion explained above.

Since the present HXD/PIN measurement gives $S = 2.34 \times 10^{35}$ erg s^{-1} (PIN FOV)$^{-1}$, the total stellar mass contained in the HXD/PIN field of view is estimated to be

$$M \sim \frac{2.34 \times 10^{35}\ \mathrm{erg\ s^{-1}\ (PIN\ FOV)^{-1}}}{0.76 \times 3.5 \times 10^{27}\ \mathrm{erg\ s^{-1}}\ M_\odot^{-1}} = 8.8 \times 10^7\ M_\odot\ \mathrm{(PIN\ FOV)^{-1}}. \tag{6.7}$$

Hereafter, we neglect errors associated with the derived values, because they could change resulting values only by 10–20 %. As reviewed in Sect. 2.2.1, WDs occupy about 10 % of the stellar mass. Therefore, the total WD mass in the HXD/PIN field of view is

$$M_{\mathrm{WD,total}} \sim 0.1M = 8.8 \times 10^6\ M_\odot\ (\mathrm{PIN\ FOV})^{-1}, \qquad (6.8)$$

and the WD number density N_{WD} can be calculated by assuming the average WD mass of 0.6 $M_\odot$ (Kepler et al. 2007) as

$$N_{\mathrm{WD}} \sim M_{\mathrm{WD,total}}/0.6\ M_\odot = 1.5 \times 10^7 (\mathrm{PIN\ FOV})^{-1}. \qquad (6.9)$$

To estimate the number densities of accreting WDs (CVs), it is required to multiply the CV/WD ratio to N_{WD}. However, this is one of the most ambiguous parameters in the CV studies, and its accurate value is still unknown spreading over a range of 10^{-3}–10^{-2} based on many measurements (for recent one, see Pretorius et al. 2007). We reluctantly assume that the ratio should be in this range, and then, the CV surface density N_{CV} is obtained as

$$N_{\mathrm{CV}} \sim 1.5 \times 10^4 - 1.5 \times 10^5 (\mathrm{PIN\ FOV})^{-1}. \qquad (6.10)$$

Finally, by multiplying the ratio of IP/CV which is about 5–10 % (Sect. 2.2), the surface number density of the hard X-ray emitting magnetic CVs (i.e., IPs) is derived

$$N_{\mathrm{IP}} \sim 7.5 \times 10^2 - 1.5 \times 10^4 (\mathrm{PIN\ FOV})^{-1}. \qquad (6.11)$$

This range consistently includes the source density calculated from the luminosity function. Based on these values, we are convinced that the seemingly high source density ($\sim$a few $\times\ 10^4$ per the HXD/PIN field of view of 1156 arcmin^{-2}) is quite normal (not so high), and naturally explained from the densities of their super classes, such as CVs and WDs.

6.3.4 The Origin of the GRXE

The wide-band spectral decomposition has been anticipated for a long time to aid in understanding the emission mechanism(s) and the energy supplier(s) of the GRXE. This has been done in the present study, decomposing the GRXE into two representative constituents. As was detailed in Sect. 6.2.3, the low-temperature and the high-temperature CIE plasma emissions have plasma temperatures of $\sim$1.0–1.5 keV and >10 keV according to our modeling. The latter component can be successfully replaced with the IP spectral model as examined in the Model 2 fits. Our spectral fit result, as explained in the previous subsection, provides even stronger support for the "Point Source" scenario of the GRXE alongside of the imaging decomposition of Revnivtsev et al. (2009).

On the other hand, the results contradict several studies that suggested that the "Diffuse" scenario was the GRXE origin. The putative non-equilibrium ionization (NEI) plasma proposed by Kaneda et al. (1997) was again clearly rejected as the GRXE origin because we detected an intense Fe XXVI Kα line in the present sky region which is not expected from their NEI model. This is a confirmation of a similar denial argument by Ebisawa et al. (2008) in the different sky region. Unlike Yamasaki et al. (1997) and Valinia and Marshall (1998), the observed wide-band spectrum did not require a putative hard-tail component which, in those literatures, was suggested to smoothly connect to the gamma-ray band up to hundreds of MeV.

6.3.5 The Mean WD Mass in the Galaxy

The WD mass of CVs is an important parameter when interpreting the hard X-ray GRXE spectral shape. It was first estimated to be $\sim$0.5 $M_\odot$ in Krivonos et al. (2007) by roughly fitting their GRXE spectra with the IP model by Suleimanov et al. (2005) (although no error was associated, models with 0.3 and 1.0 $M_\odot$ did not fit their data according to their Fig. 16). Our hard X-ray analysis confirmed this, and gave a slightly heavier WD mass of $0.66^{+0.09}_{-0.07}$ $M_\odot$. As Krivonos et al. (2007) mentioned, the value could be interpreted as the mean WD mass in magnetic CVs in the Galaxy.

If we accept this idea, it is intriguing to compare the derived WD mass with that reported for isolated WDs based on optical spectroscopy. For example, using large SDSS data, Kepler et al. (2007) reported a mean WD mass of 0.593 ± 0.016 $M_\odot$ for 1733 WDs that have helium and hydrogen outer layers. Our value $0.66^{+0.09}_{-0.07}$ $M_\odot$ includes this within its rather large errors. The slight difference of the center values ($\sim$0.06 $M_\odot$) could be a result of long-lasting mass accretion; rates of the accretion taking place in CVs are $\dot{M} \sim 10^{-11}$–10^{-9} $M_\odot$ (e.g., our measurement listed in Table 5.5), and the mass increase may occur after $\sim$$10^9$–$10^7$ years of continuous accretion. However, CVs are thought to evolve through the common-envelope phase of main-sequence stars, in which two stars in a binary share their outer layers, possibly resulting in WD masses different from those of isolated stars. Therefore, at this moment, it is difficult to investigate the difference of the average WD masses. More accurate WD mass determinations in CVs, and a sophisticated evolution model of them are necessary to enable this.

6.3.6 Connection to the Galactic Center X-ray Emission

A low-temperature CIE component ($kT \sim 1$ keV) similar to one seen in the GRXE was also reported recently by Nobukawa et al. (2010) in the apparently extended X-ray emission near the Galactic center region ($|l| < 0.2°$). Probably due to a lack of hard X-ray spectral coverage in their analysis (used only the 2–10 keV range), Nobukawa et al. (2010) reported a plasma temperature of the hotter CIE component of 7.0 ± 0.1 keV, which is significantly cooler than our result $kT = 15.1^{+0.4}_{-0.7}$ keV. Note also that their spectral analysis introduced, in addition to two CIE plasma components, an intense power-law component which was not detected in our wide-band GRXE data.

In the Galactic center region, the supermassive black hole ($\sim 3 \times 10^6\ M_\odot$) is suspected to have a close connection with the unresolved (diffuse) emission which is observed in its proximity (e.g., predicted and observed by Sunyaev et al. 1993 and Inui et al. 2009). This emission could have another origin which differs from that of the GRXE. Besides, we stress that the stellar density is higher by more than an order of magnitude in the very central region of the Galaxy ($\sim$100 pc from the Galactic center) compared to those of the region we studied. Therefore, X-rays from faint point sources (mostly active binaries and IPs) might occupy significant fraction of the detected signals, although Nobukawa et al. (2010) simply neglected them. Since about a half of the faint point sources radiates their energy in the hard X-ray band, it is now obvious that a wide-band spectral analysis including the hard X-ray energy range is essential to examine possible existence of truly diffuse emission.

6.4 Summary of the GRXE Study

We accumulated the data from the *Suzaku* GRXE observations in the Galactic bulge region achieving $\sim$1 Ms exposure in total (Sect. 6.1.1), and produced the wide-band GRXE spectra with high counting statistics in 2–50 keV band (Sect. 6.1.5). Because of the careful data selection and the tightly collimated field of view of HXD/PIN, we achieved the ever lowest limiting flux for a contaminating bright point source among wide-band GRXE studies (0.1 mCrab, equivalently $\sim 10^{12}$ erg cm^{-2} s^{-1} and in the 2–8 keV; Sect. 6.1.2).

The XIS and HXD/PIN spectra were separately analyzed. We revealed that the hard X-ray GRXE spectrum taken with HXD/PIN is reproduced well with the IP model (Sect. 6.2.1) with a WD mass of $0.66^{+0.09}_{-0.07}\ M_\odot$. From emission line analyses based on CIE modeling which is more appropriate than previous line studies (Kaneda et al. 1997), we reconfirmed the multi-temperature nature of the GRXE (Sect. 6.2.2).

Based on these results, we constructed physical models of the GRXE (Sect. 6.2.3) which consists of multi-temperature CIE plasma emissions. The models nicely reproduced the wide-band GRXE spectral shape for the first time. In particular, one including the IP component (Model 2) gave better fits to the data. WD masses of 0.66 $M_\odot$ (fixed) or $0.48^{+0.05}_{-0.04}\ M_\odot$ reproduced the GRXE spectral shape with almost equivalent fitting statistics, together with the CIE plasma component with $kT \sim 1.0$–1.6 keV. These plasma temperature and the WD mass(es) are quite consistent with those of active binaries and magnetic accreting WDs (Sect. 6.3.1).

We further calculated the X-ray luminosity function and the number densities of the unresolved hard X-ray point sources, mostly IPs (Sect. 6.3.3). The luminosity function is consistent with those obtained with a deep imaging observation and individual source counting in the solar vicinity, indicating that no special type of X-ray source is required to explain the GRXE flux. The source density (~a few × 10^4 in the observed area, 1156 arcmin^2) is beautifully consistent with that of IPs independently estimated from the Galactic stellar population.

Combining the spectral decomposition result and the calculated source density, we conclude that the present result strongly supports the "Point Source" scenario of the GRXE origin.

References

Andrievsky, S. M., Bersier, D., Kovtyukh, V. V., et al. 2002a, A&A, 384, 140
Andrievsky, S. M., Kovtyukh, V. V., Luck, R. E., et al. 2002b, A&A, 381, 32
Blay, P., Martínez-Núñez, S., Negueruela, I., et al. 2008, A&A, 489, 669
Bozzo, E., Stella, L., Ferrigno, C., et al. 2010, A&A, 519, A6
Bryans, P., Badnell, N. R., Gorczyca, T. W., et al. 2006, ApJS, 167, 343
Bryans, P., Landi, E., & Savin, D. W. 2009, ApJ, 691, 1540
Covino, S., Tagliaferri, G., Pallavicini, R., Mewe, R., & Poretti, E. 2000, A&A, 355, 681
Dwek, E., Arendt, R. G., Hauser, M. G., et al. 1995, ApJ, 445, 716
Ebisawa, K., Bourban, G., Bodaghee, A., Mowlavi, N., & Courvoisier, T. 2003, A&A, 411, L59
Ebisawa, K., Tsujimoto, M., Paizis, A., et al. 2005, ApJ, 635, 214
Ebisawa, K., Yamauchi, S., Tanaka, Y., et al. 2008, PASJ, 60, 223
Favata, F., Mewe, R., Brickhouse, N. S., et al. 1997, A&A, 324, L37
Fukazawa, Y., Mizuno, T., Watanabe, S., et al. 2009, PASJ, 61, 17
Grindlay, J. E., Covault, C. E., & Manandhar, R. P. 1993, A&AS, 97, 155
Harrison, F. A., Boggs, S., Christensen, F., et al. 2010, in the Society of Photo-Optical Instrumentation Engineers (SPIE) Conference, Vol. 7732, SPIE Conference Series
Inui, T., Koyama, K., Matsumoto, H., & Tsuru, T. G. 2009, PASJ, 61, 241
Kalberla, P. M. W., Burton, W. B., Hartmann, D., et al. 2005, A&A, 440, 775
Kaneda, H., Makishima, K., Yamauchi, S., et al. 1997, ApJ, 491, 638
Kepler, S. O., Kleinman, S. J., Nitta, A., et al. 2007, MNRAS, 375, 1315
Kimura, S. 2010, PhD thesis, The University of Tokyo
Kinugasa, K. & Tsunemi, H. 1999, PASJ, 51, 239
Krivonos, R., Revnivtsev, M., Churazov, E., et al. 2007, A&A, 463, 957
Kuntz, K. D. & Snowden, S. L. 2008, ApJ, 674, 209
Masui, K., Mitsuda, K., Yamasaki, N. Y., et al. 2009, PASJ, 61, 115
Muno, M. P., Arabadjis, J. S., Baganoff, F. K., et al. 2004, ApJ, 613, 1179
Muno, M. P., Bauer, F. E., Baganoff, F. K., et al. 2009, ApJS, 181, 110
Nobukawa, M., Koyama, K., Tsuru, T. G., Ryu, S. G., & Tatischeff, V. 2010, PASJ, 62, 423
Nobukawa, M., Tsuru, T. G., Takikawa, Y., et al. 2008, PASJ, 60, 191
Pandel, D., Córdova, F. A., Mason, K. O., & Priedhorsky, W. C. 2005, ApJ, 626, 396
Pretorius, M. L., Knigge, C., O'Donoghue, D., et al. 2007, MNRAS, 382, 1279
Revnivtsev, M., Sazonov, S., Churazov, E., et al. 2009, Nature, 458, 1142
Revnivtsev, M., Sazonov, S., Gilfanov, M., Churazov, E., & Sunyaev, R. 2006, A&A, 452, 169
Revnivtsev, M., Sazonov, S., Krivonos, R., Ritter, H., & Sunyaev, R. 2008, A&A, 489, 1121
Sakano, M., Koyama, K., Murakami, H., Maeda, Y., & Yamauchi, S. 2002, ApJS, 138, 19
Sawada, M., Tsujimoto, M., Koyama, K., et al. 2009, PASJ, 61, 209
Sazonov, S., Revnivtsev, M., Gilfanov, M., Churazov, E., & Sunyaev, R. 2006, A&A, 450, 117
Sidoli, L., Romano, P., Ducci, L., et al. 2009, MNRAS, 397, 1528
Suleimanov, V., Revnivtsev, M., & Ritter, H. 2005, A&A, 435, 191
Sunyaev, R. A., Markevitch, M., & Pavlinsky, M. 1993, ApJ, 407, 606

Takahashi, T., Kelley, R., Mitsuda, K., et al. 2008, in SPIE Conference Series, Vol. 7011
Tamagawa, T., Hayato, A., Nakamura, S., et al. 2009, PASJ, 61, 167
Tanaka, Y. 2002, A&A, 382, 1052
Tawa, N., Hayashida, K., Nagai, M., et al. 2008, PASJ, 60, 11
Türler, M., Chernyakova, M., Courvoisier, T., et al. 2010, A&A, 512, A49
Uchiyama, H. 2010, PhD thesis, Kyoto University
Valinia, A. & Marshall, F. E. 1998, ApJ, 505, 134
Valinia, A., Tatischeff, V., Arnaud, K., Ebisawa, K., & Ramaty, R. 2000, ApJ, 543, 733
Yamasaki, N. Y., Ohashi, T., Takahara, F., et al. 1997, ApJ, 481, 821
Yoshino, T., Mitsuda, K., Yamasaki, N. Y., et al. 2009, PASJ, 61, 805

Chapter 7
Conclusion

To spectroscopically study the origin of the Galactic X-ray background emission, or so-called Galactic Ridge X-ray Emission (GRXE), we analyzed individual X-ray spectra of 17 nearby Intermediate Polars (IPs) observed with the *Suzaku* satellite. This is because IPs are thought to be a major contributor to the GRXE in the hard X-ray energy band above $E > 10\,\mathrm{keV}$. Using the numerically constructed spectral model, we fitted their spectra for the first time simultaneously covering the two important spectral characteristics, namely the resolved Fe emission lines and the thermal cutoff (roll-off) of the continuum in the hard X-ray band. The models gave WD mass estimates in each source, with typical fitting errors of 0.05–0.15 $M_{\odot}$. The mean WD mass is $0.88 \pm 0.23\,M_{\odot}$ for the present sample.

Following the individual IP studies, we produced wide-band GRXE spectra covering the 2–50 keV band, by accumulating the *Suzaku* GRXE observations in two sky regions. Thanks to the very deep exposure ($\sim$1 Ms), the spectra had high counting statistics, and resolved emission lines from highly ionized Si, S, Ar, and Fe ions and neutral Fe. Through careful examination of the data as well as physical considerations, we arrived at a GRXE spectral model which consists of a single-temperature CIE plasma emission and the IP spectral model. The composite model successfully reproduced the observed wide-band GRXE spectra, with an implication that the GRXE is explicable as an assembly of numerous IPs and active binary stars, contributing in the high and low energy bands, respectively. This provides the first quantitative wide-band spectral decomposition of the GRXE, together with a number density estimate of hard X-ray sources required to fulfill the observed surface brightness.

The past and coming five years are the golden age of the GRXE study. The present spectral decomposition and results of the *Chandra* deep imaging studies (Revnivtsev and Sazonov 2007; Revnivtsev et al. 2009) consistently support the "Point Source" scenario for the origin of the GRXE from two definitive and complementary aspects, i.e., imaging and wide-band spectroscopic decompositions. After more than forty years from its discovery, we have eventually clarified that the Galaxy is filled with numerous low-luminosity X-ray sources, radiating via magnetic activities and liberation of gravitational potential energy.

T. Yuasa, Suzaku *Studies of White Dwarf Stars and the Galactic X-ray Background Emission*, Springer Theses, DOI: 10.1007/978-4-431-54219-3_7, © Springer Japan 2013

In early 2014, the *ASTRO-H* satellite (Takahashi et al. 2008) will allow us to disclose detailed properties of individual contributors of the GRXE by completely resolving the fine structures of the atomic emission lines and by determining the hard X-ray spectral shape much more precisely.

References

Revnivtsev, M. & Sazonov, S. 2007, A&A, 471, 159
Revnivtsev, M., Sazonov, S., Churazov, E., et al. 2009, Nature, 458, 1142
Takahashi, T., Kelley, R., Mitsuda, K., et al. 2008, in SPIE Conference Series, Vol. 7011

Appendix A

A.1 Spectral Fitting Results of Individual Sources

In Fig. A.1, we present the observed spectra of Intermediate Polars analyzed in Chap. 5 overlaid with the best-fit model spectra.

T. Yuasa, Suzaku *Studies of White Dwarf Stars and the Galactic X-ray Background Emission,* Springer Theses DOI: 10.1007/978-4-431-54219-3, © Springer Japan 2013

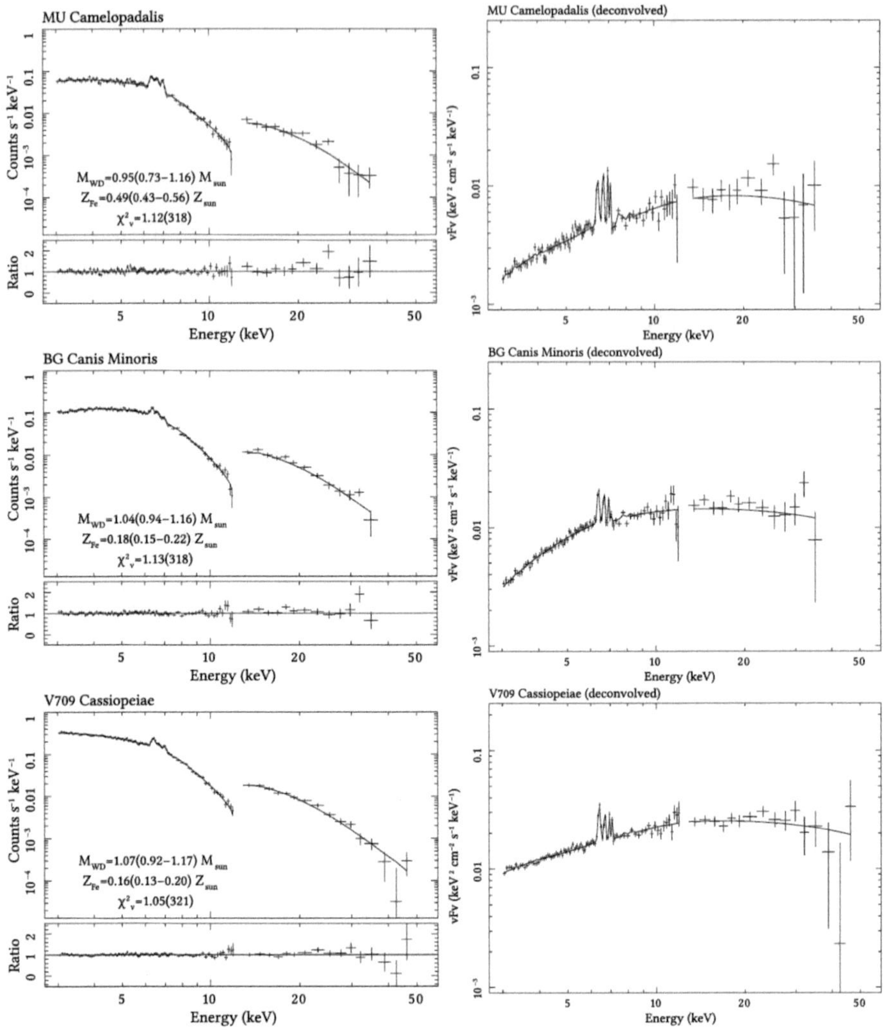

Fig. A.1 Continued from Fig. 5.7

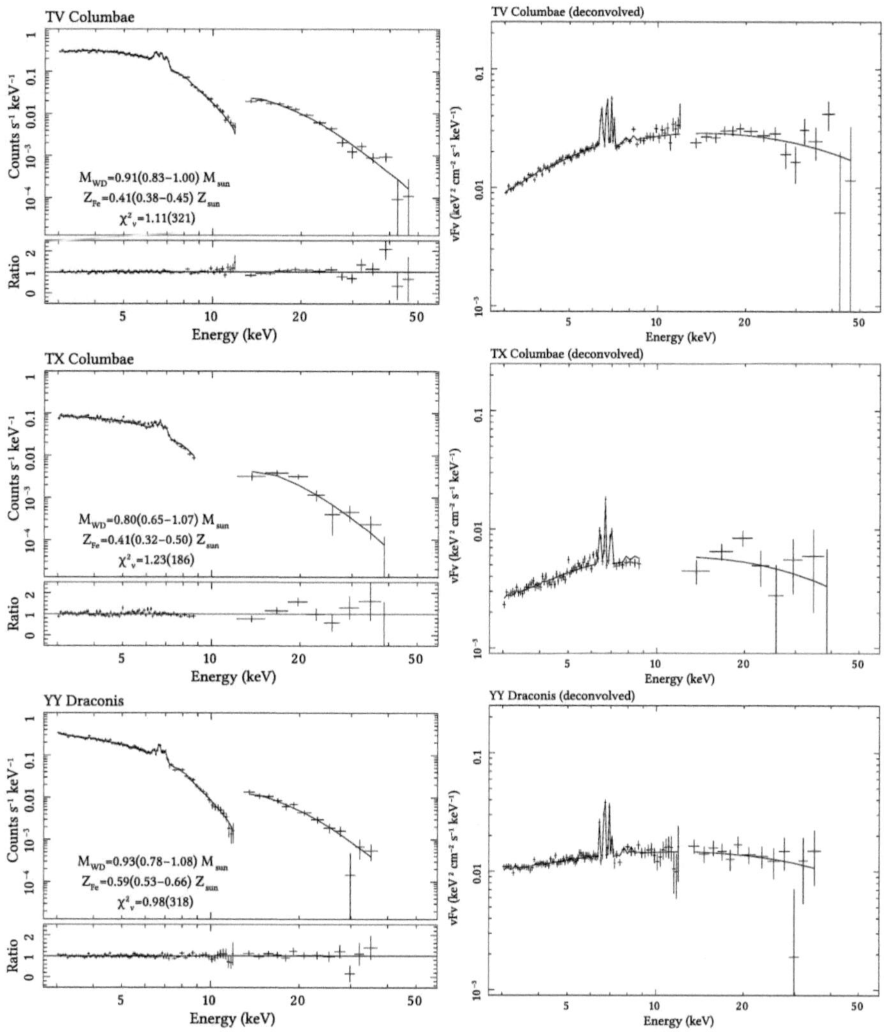

Fig. A.1 (continued)

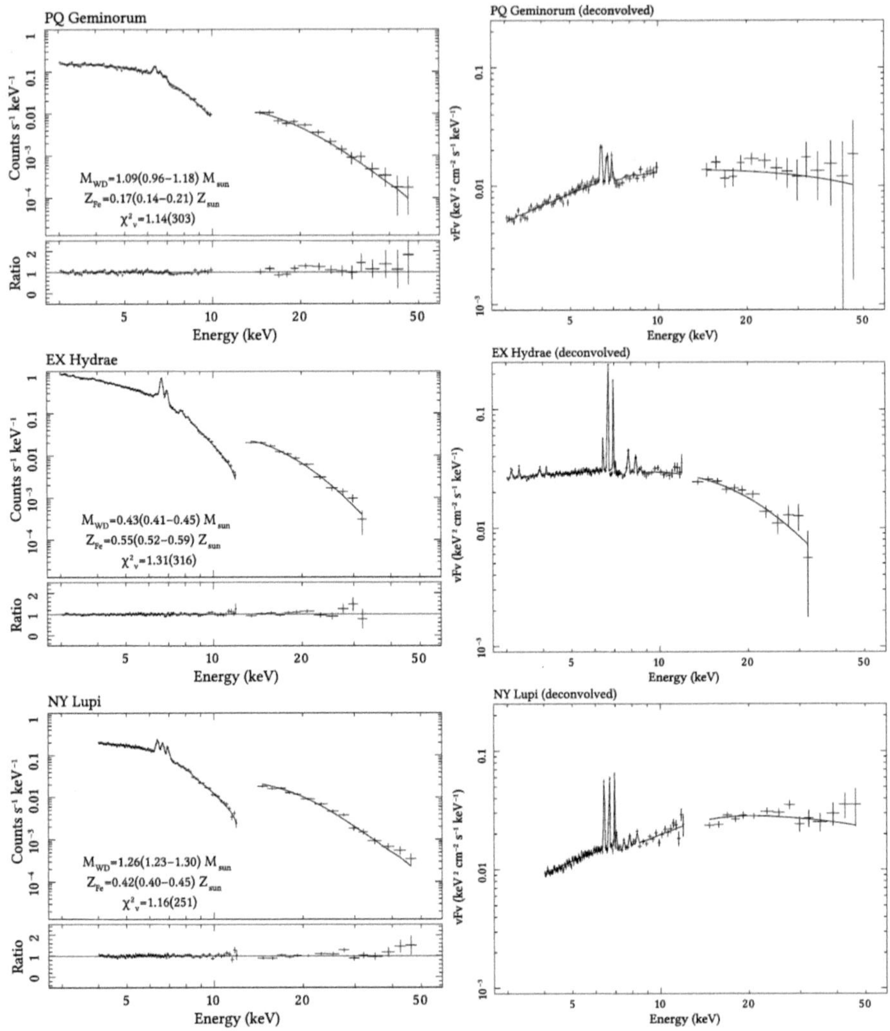

Fig. A.1 (continued)

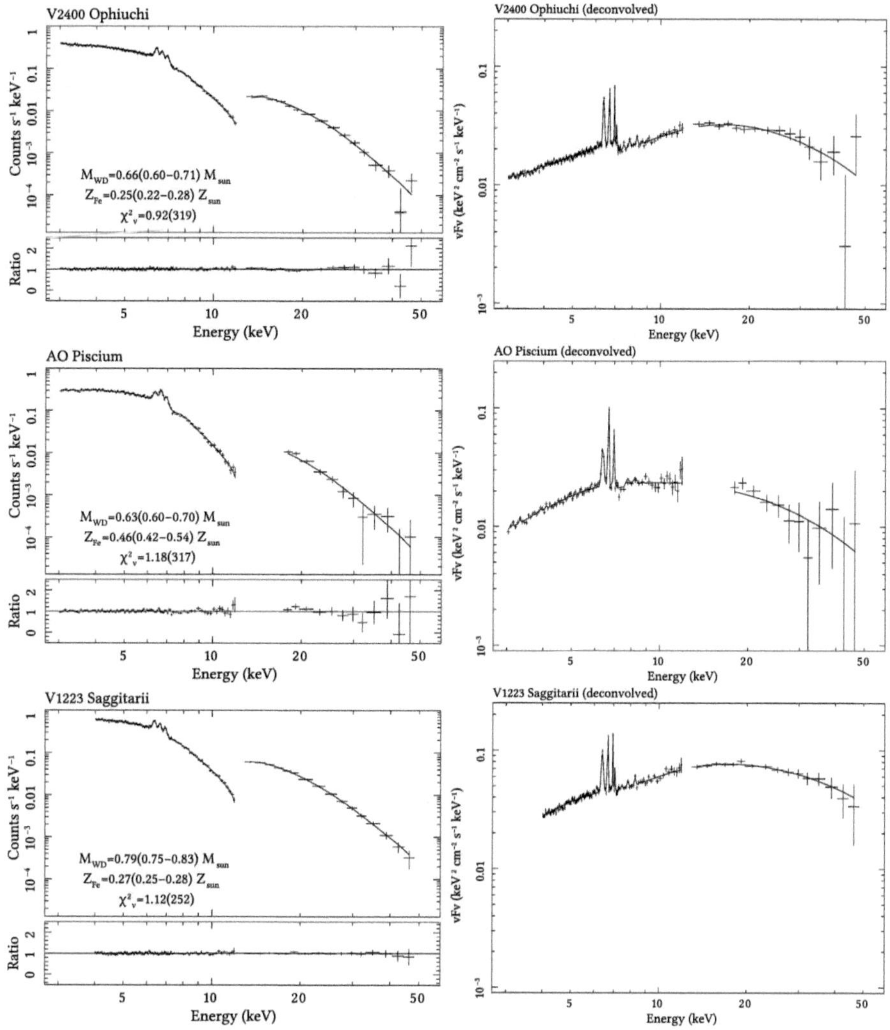

Fig. A.1 (continued)

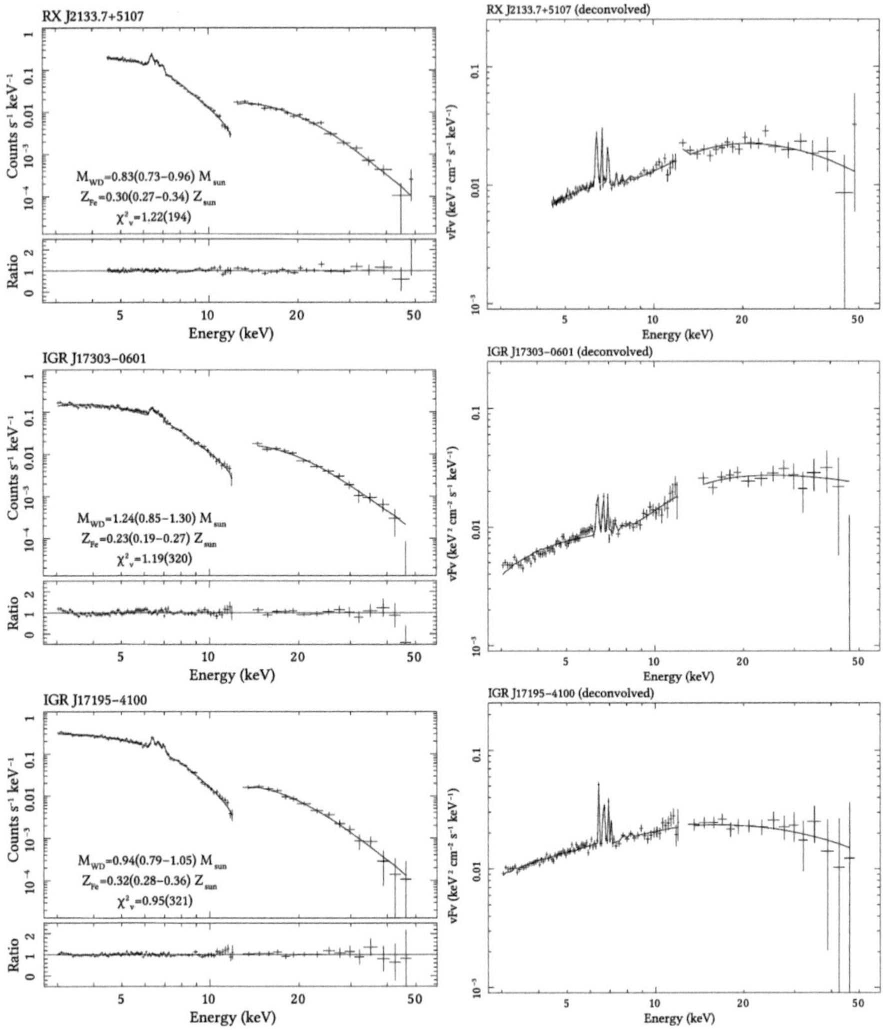

Fig. A.1 (continued)

Appendix B

B.1 *Suzaku* Observations of the Galactic Center

For later reference, we summarize the data from 92 *Suzaku* observations in the Galactic center region in Table B.1. XIS data of these observations are used to produce the mosaic image shown in Fig. 6.1. The total effective exposures of the XIS and HXD/PIN are 4.4 and 3.9 Ms. The list contains observations with various types of scientific aims: the Galactic Ridge X-ray Emission, supernova remnants, transient X-ray binaries, molecular clouds reflecting X-rays, and unidentified sources in TeV wavelength (H.E.S.S unID sources).

T. Yuasa, Suzaku *Studies of White Dwarf Stars and the Galactic X-ray Background Emission,* Springer Theses DOI: 10.1007/978-4-431-54219-3, © Springer Japan 2013

Table B.1 *Suzaku* observations of the Galactic center

	Obs. ID[a]	Coordinate[b]		Start time	Exposure[c]		P.I.
		l	b	UT	XIS	PIN	
1	100027010	0.06	−0.08	2005-09-23 07:18:25	44.8	37.9	Suzaku Team
2	100027020	−0.24	−0.05	2005-09-24 14:17:17	42.8	36.1	Suzaku Team
3	100027030	−0.44	−0.39	2005-09-24 11:07:08	2.1	1.9	Suzaku Team
4	100027040	−0.44	−0.07	2005-09-24 12:41:33	1.9	1.8	Suzaku Team
5	100027050	0.33	0.01	2005-09-25 17:29:12	2.0	1.8	Suzaku Team
6	100037010	−0.24	−0.05	2005-09-29 04:35:41	43.7	39.4	Suzaku Team
7	100037020	−0.44	−0.39	2005-09-30 04:30:44	3.3	3.1	Suzaku Team
8	100037030	−0.45	−0.07	2005-09-30 06:06:32	3.0	2.8	Suzaku Team
9	100037040	0.06	−0.08	2005-09-30 07:43:01	43.0	39.5	Suzaku Team
10	100037050	0.33	0.01	2005-10-01 06:22:41	2.4	2.2	Suzaku Team
11	100037060	0.64	−0.10	2005-10-10 12:28:01	76.6	70.8	Suzaku Team
12	100037070	1.00	−0.10	2005-10-12 07:10:24	9.2	9.5	Suzaku Team
13	100048010	0.06	−0.08	2006-09-08 02:23:24	63.0	60.3	Suzaku Team
14	102013010	0.06	−0.08	2007-09-03 19:01:10	51.4	44.5	Hironori Matsumoto
15	402066010	−1.93	0.45	2008-02-22 11:52:49	36.5	31.3	John A. Tomsick
16	403001010	1.36	1.05	2009-02-22 19:04:19	71.5	59.7	Junichiro Miura
17	403009010	0.17	0.03	2009-03-21 02:03:28	110.8	91.7	Yoshiaki Hyodo
18	500005010	0.43	−0.11	2006-03-27 23:00:22	88.4	64.6	Suzaku Team
19	500018010	−0.57	−0.09	2006-02-20 12:45:25	106.9	46.6	Suzaku Team
20	500019010	−1.09	−0.04	2006-02-23 10:51:11	13.3	12.2	Suzaku Team
21	501008010	−0.16	−0.19	2006-09-2614:18:16	129.6	111.3	Katsuji Koyama
22	501009010	−0.07	0.18	2006-09-29 21:26:07	51.2	47.7	Katsuji Koyama
23	501010010	−1.29	−0.64	2006-10-07 02:16:52	50.7	45.7	Ryo Yamazaki
24	501039010	0.78	−0.16	2007-03-03 12:20:20	96.4	91.1	Katsuji Koyama
25	501040010	0.61	0.07	2006-09-21 17:29:01	61.4	53.9	Katsuji Koyama
26	501040020	0.61	0.07	2006-09-24 05:03:12	44.8	40.0	Katsuji Koyama
27	501046010	−0.17	0.34	2007-03-10 15:03:10	25.2	25.0	Katsuji Koyama
28	501047010	−0.50	0.34	2007-03-11 03:55:59	25.6	19.1	Katsuji Koyama
29	501048010	−0.83	0.34	2007-03-11 19:04:59	27.5	24.1	Katsuji Koyama
30	501049010	−1.17	0.33	2006-10-08 10:22:40	19.6	17.6	Katsuji Koyama
31	501050010	−0.83	−0.00	2006-10-09 02:20:25	22.0	18.6	Katsuji Koyama
32	501051010	−1.17	−0.00	2006-10-09 13:40:09	21.9	21.1	Katsuji Koyama
33	501052010	−1.50	−0.00	2006-10-10 06:45:09	19.3	16.0	Katsuji Koyama
34	501053010	−1.83	−0.00	2006-10-10 21:18:59	21.9	19.9	Katsuji Koyama
35	501054010	−0.17	−0.33	2007-03-12 08:11:07	26.1	23.5	Katsuji Koyama
36	501055010	−0.50	−0.33	2007-03-12 23:59:09	27.2	21.2	Katsuji Koyama
37	501056010	−0.83	−0.33	2007-03-13 15:41:12	26.5	25.3	Katsuji Koyama
38	501057010	−1.17	−0.34	2006-10-11 10:07:27	20.5	19.1	Katsuji Koyama
39	501058010	1.30	0.20	2007-03-14 05:02:29	63.3	51.1	Katsuji Koyama
40	501059010	1.17	0.00	2007-03-15 18:55:51	62.2	54.4	Katsuji Koyama
41	501060010	1.50	0.00	2007-03-17 05:07:04	64.8	54.6	Katsuji Koyama
42	502002010	0.17	−0.67	2007-10-09 16:40:54	23.2	20.9	Motohide Kokubun
43	502003010	−0.17	−0.67	2007-10-10 03:41:13	21.5	18.9	Motohide Kokubun

(continued)

Table B.1 (continued)

	Obs. ID[a]	Coordinate[b]		Start time	Exposure[c]		P.I.
		l	*b*	UT	XIS	PIN	
44	502004010	0.17	−1.00	2007-10-10 15:21:17	19.9	18.8	Motohide Kokubun
45	502005010	−0.17	−1.00	2007-10-11 01:01:17	20.6	18.2	Motohide Kokubun
46	502006010	0.17	0.33	2007-10-11 11:34:01	22.6	21.7	Motohide Kokubun
47	502007010	0.17	0.66	2007-10-11 23:09:15	22.0	19.5	Motohide Kokubun
48	502008010	−0.17	0.66	2007-10-12 09:52:59	23.8	22.9	Motohide Kokubun
49	502009010	1.83	−0.00	2007-10-12 21:52:24	20.9	19.6	Motohide Kokubun
50	502010010	0.50	0.33	2007-10-13 07:32:00	21.6	21.2	Motohide Kokubun
51	502011010	0.83	0.33	2007-10-13 18:51:09	23.0	22.1	Motohide Kokubun
52	502016010	−1.08	−0.48	2008-03-02 18:08:00	70.5	61.8	Ryo Yamazaki
53	502017010	−0.95	−0.65	2008-03-06 13:26:36	72.6	64.0	Ryo Yamazaki
54	502018010	−1.27	−0.42	2008-03-08 16:02:17	79.0	70.2	Ryo Yamazaki
55	502020010	1.05	−0.17	2007-09-06 00:26:47	139.1	124.5	Katsuji Koyama
56	502022010	0.23	−0.27	2007-08-31 12:33:33	134.8	116.8	Takeshi Go Tsuru
57	502051010	0.92	0.01	2008-03-11 06:19:45	138.8	122.2	Katsuji Koyama
58	502059010	−0.00	−2.00	2007-09-29 01:40:51	136.8	110.5	Motohide Kokubun
59	503007010	0.33	0.17	2008-09-02 10:15:27	52.2	44.2	Katsuji Koyama
60	503008010	0.00	−0.38	2008-09-03 22:53:29	53.7	42.8	Katsuji Koyama
61	503009010	−0.32	−0.24	2008-09-05 06:57:08	52.4	40.3	Katsuji Koyama
62	503010010	−0.69	−0.05	2008-09-06 15:56:13	53.1	37.1	Katsuji Koyama
63	503011010	−0.97	−0.13	2008-09-08 09:08:09	57.6	40.2	Katsuji Koyama
64	503012010	−0.91	−0.45	2008-09-14 19:35:07	57.7	51.9	Katsuji Koyama
65	503013010	−1.30	−0.05	2008-09-16 00:51:19	104.8	93.9	Katsuji Koyama
66	503014010	−2.10	−0.05	2008-09-18 04:46:49	55.4	51.2	Katsuji Koyama
67	503015010	−2.35	−0.05	2008-09-19 07:33:05	56.8	52.8	Katsuji Koyama
68	503016010	−2.60	−0.05	2008-09-22 06:47:49	52.2	49.3	Katsuji Koyama
69	503017010	−2.85	−0.05	2008-09-23 08:08:10	51.3	48.6	Katsuji Koyama
70	503021010	−1.62	0.20	2008-10-04 03:44:03	53.8	49.6	Katsuji Koyama
71	503072010	−0.42	0.17	2009-03-06 02:39:12	140.6	135.5	Takeshi Go Tsuru
72	503076010	−1.50	0.15	2009-02-24 17:04:51	52.9	43.8	Hironori Matsumoto
73	503077010	−1.70	0.14	2009-02-26 01:01:00	51.3	43.7	Hironori Matsumoto
74	503081010	0.03	−1.66	2009-03-09 15:41:50	59.2	57.6	Hiroshi Murakami
75	503099010	−0.22	1.13	2009-03-10 19:39:08	29.7	30.6	Farhad Yusef-Zadeh
76	503100010	−0.69	1.13	2009-03-15 06:41:41	25.7	24.1	Farhad Yusef-Zadeh
77	503101010	−0.45	0.89	2009-03-16 14:43:17	33.9	30.8	Farhad Yusef-Zadeh
78	503102010	−0.70	0.66	2009-03-17 07:49:09	33.7	30.1	Farhad Yusef-Zadeh
79	503103010	−0.01	1.20	2009-03-11 10:56:59	18.3	16.4	Farhad Yusef-Zadeh
80	504050010	0.10	−1.42	2010-03-06 03:55:37	100.4	80.5	Ken Ebisawa
81	504088010	−0.00	−0.83	2009-10-14 11:30:56	47.2	32.6	Katsuji Koyama
82	504089010	−0.05	−1.20	2009-10-09 04:05:59	55.3	40.2	Katsuji Koyama
83	504090010	−1.49	−1.18	2009-10-13 04:17:20	41.3	35.0	Katsuji Koyama
84	504091010	−1.50	−1.60	2009-09-14 19:37:36	51.3	47.8	Katsuji Koyama
85	504092010	−1.44	−2.15	2009-09-16 07:21:35	50.9	45.6	Katsuji Koyama
86	504093010	−1.50	−2.80	2009-09-17 13:54:31	53.2	46.9	Katsuji Koyama

(continued)

Table B.1 (continued)

	Obs. ID[a]	Coordinate[b]		Start time	Exposure[c]		P.I.
		l	b	UT	XIS	PIN	
87	903004010	−2.75	−1.84	2008-10-07 16:19:21	15.7	28.7	Suzaku Team
88	904002010	1.02	2.53	2009-08-28 12:20:31	23.1	21.9	Suzaku Team
89	904002020	1.02	2.53	2009-09-06 19:38:32	25.1	18.8	Suzaku Team
90	504003010	−1.45	−0.87	2010-02-25 04:33:17	50.9	41.3	Katsuji Koyama
91	504001010	−1.47	−0.26	2010-02-26 09:15:00	51.2	42.2	Katsuji Koyama
92	504002010	−1.53	−0.58	2010-02-27 16:14:41	53.1	46.6	Katsuji Koyama

[a]Observation ID

[b]Aim point in Galactic coordinate (degree)

[c]Net exposure in units of 10^3 s

Appendix C

C.1 X-Ray Spectra of Dwarf Novae

Suzaku has been observing several dwarf novae as listed in Table C.1. Although the listed *Suzaku* observations of dwarf novae are not complete in any sense but rather heterogeneous because individual observations were separately proposed for different scientific purposes, their general properties will be helpful understanding dwarf novae contribution to the Galactic ridge X-ray emission (GRXE). Therefore, we briefly study these objects, and report spectral characteristics (e.g. plasma temperature) based on model fits (Table C.2).

Pandel et al. (2005) showed that X-ray spectra of dwarf novae at low accretion rates (i.e. quiescent states) can be well reproduced with a thermal emission from a cooling plasma (plasma with temperature gradient). Following this study, we also fitted dwarf novae spectra obtained with *Suzaku*, first using a single-temperature CIE plasma emission model (APEC model in XSPEC), and then the IP model constructed in the present thesis. The IP model is also a cooling plasma model which can mimic dwarf novae spectra although it is not equivalent to the model used in Pandel et al. (2005).

As shown in Figs. C.1 and C.2, spectra of the observed dwarf novae well reproduced by both the models. In the fits, the plasma emission models were subjected to the interstellar absorption assuming hydrogen column densities measured toward the targets (Kalberla et al. 2005). In addition to the hot plasma component, i.e. the single-temperature CIE or the IP model, a gaussian function was introduced in the model to fulfill Fe I Kα line which is probably produced via reflection of X-rays by neutral Fe in the accretion disc.

The best-fit plasma temperatures were in ranges of 4–9 keV, and no additional absorption (intrinsic absorption like IPs) was required from the data. When compared with other candidates of the GRXE origin, these temperatures are intermediate between those of magnetic cataclysmic variables (CVs) and active binaries. With this result, the reduction of the WD mass from $0.66^{+0.09}_{-0.07}\,M_{\odot}$ (Sect. 6.2.1) to $0.48^{+0.05}_{-0.04}\,M_{\odot}$ in the wide-band GRXE spectral analysis can be

T. Yuasa, Suzaku *Studies of White Dwarf Stars and the Galactic X-ray Background Emission,* Springer Theses DOI: 10.1007/978-4-431-54219-3, © Springer Japan 2013

Table C.1 *Suzaku* observations of dwarf novae

Obs.ID	Name	Coordinate[a]		Date	Exp.[b]	Rate[c]
		RA	Dec	(UT)	(ks)	
401041010	V893 SCO	243.812	−28.626	2006-08-26 09:39:29	18.5	1.37
402043010	VY AQR	318.039	−8.827	2007-11-10 15:40:07	25.4	0.05
402044010	SW UMA	129.178	53.477	2007-11-06 05:40:57	16.9	0.08
402045010	SS AUR	93.344	47.740	2008-03-04 18:01:16	19.5	0.15
402046010	BZ UMA	133.434	57.811	2008-03-24 23:49:24	29.8	0.13
403039010	ASAS J002511	6.296	12.287	2009-01-10 16:21:11	33.3	0.02
403041010	KT PER	24.287	50.956	2009-01-12 21:29:49	29.2	0.14

[a]Sky coordinate of the target in units of degree (J2000 equinox)
[b]Net exposure in 1000 s
[c]Background-subtracted XIS count rate in units of counts s^{-1}

Table C.2 The best-fit parameters obtained with the single-temperature CIE model and the IP model fits

Name	Abs.[a]	CIE Model Fit[b]			IP Model[c]		
	n_H 10^{22} cm^{-2}	kT (keV)	Z ($Z_\odot$)	χ^2_ν	M_{WD} ($M_\odot$)	Z ($Z_\odot$)	χ^2_ν
V893 SCO	0.12	$9.52^{+0.45}_{-0.45}$	$0.96^{+0.08}_{-0.07}$	1.29(505)	$1.16^{+0.04}_{-0.04}$	$1.29^{+0.12}_{-0.12}$	1.16(505)
VY AQR	0.05	$6.13^{+1.74}_{-1.08}$	$0.47^{+0.27}_{-0.23}$	1.22(34)	$0.49^{+0.18}_{-0.10}$	$0.57^{+0.31}_{-0.26}$	1.16(34)
SW UMA	0.04	$4.99^{+1.11}_{-0.75}$	$0.22^{+0.13}_{-0.18}$	1.08(34)	$0.39^{+0.10}_{-0.07}$	$0.27^{+0.17}_{-0.27}$	1.00(34)
SS AUR	0.16	$8.22^{+1.29}_{-0.85}$	$1.10^{+0.22}_{-0.20}$	0.95(73)	$0.68^{+0.51}_{-0.11}$	$1.30^{+0.13}_{-0.13}$	0.93(73)
BZ UMA	0.04	$5.99^{+0.57}_{-0.54}$	$0.60^{+0.12}_{-0.11}$	1.08(95)	$0.45^{+0.05}_{-0.05}$	$0.80^{+0.17}_{-0.14}$	1.03(95)
ASAS J002511	0.04	$4.81^{+1.68}_{-1.04}$	$0.84^{+0.41}_{-0.42}$	1.32(21)	$0.37^{+0.14}_{-0.09}$	$0.99^{+0.45}_{-0.45}$	1.27(21)
KT PER	0.14	$5.61^{+0.54}_{-0.50}$	$0.39^{+0.09}_{-0.08}$	0.98(103)	$0.43^{+0.05}_{-0.04}$	$0.50^{+0.12}_{-0.11}$	0.98(103)

[a]Absorption column density fixed at values reported by Kalberla et al. (2005); in units of 10^{22} cm^{-2}
[b]Plasma temperature and metal abundance derived with the single-temperature CIE plasma fit
[c]WD mass and metal abundance derived with the IP model fit

understood naturally, because the representative plasma temperature would decrease from those of magnetic CVs if some fraction of the GRXE flux is emitted from dwarf novae. Contribution from magnetic CVs and dwarf novae to the GRXE is 1.2:0.6 based on the fractional contributions to the GRXE predicted by Revnivtsev et al. (2006) and Sazonov et al. (2006), and thus the above discussion seems feasible.

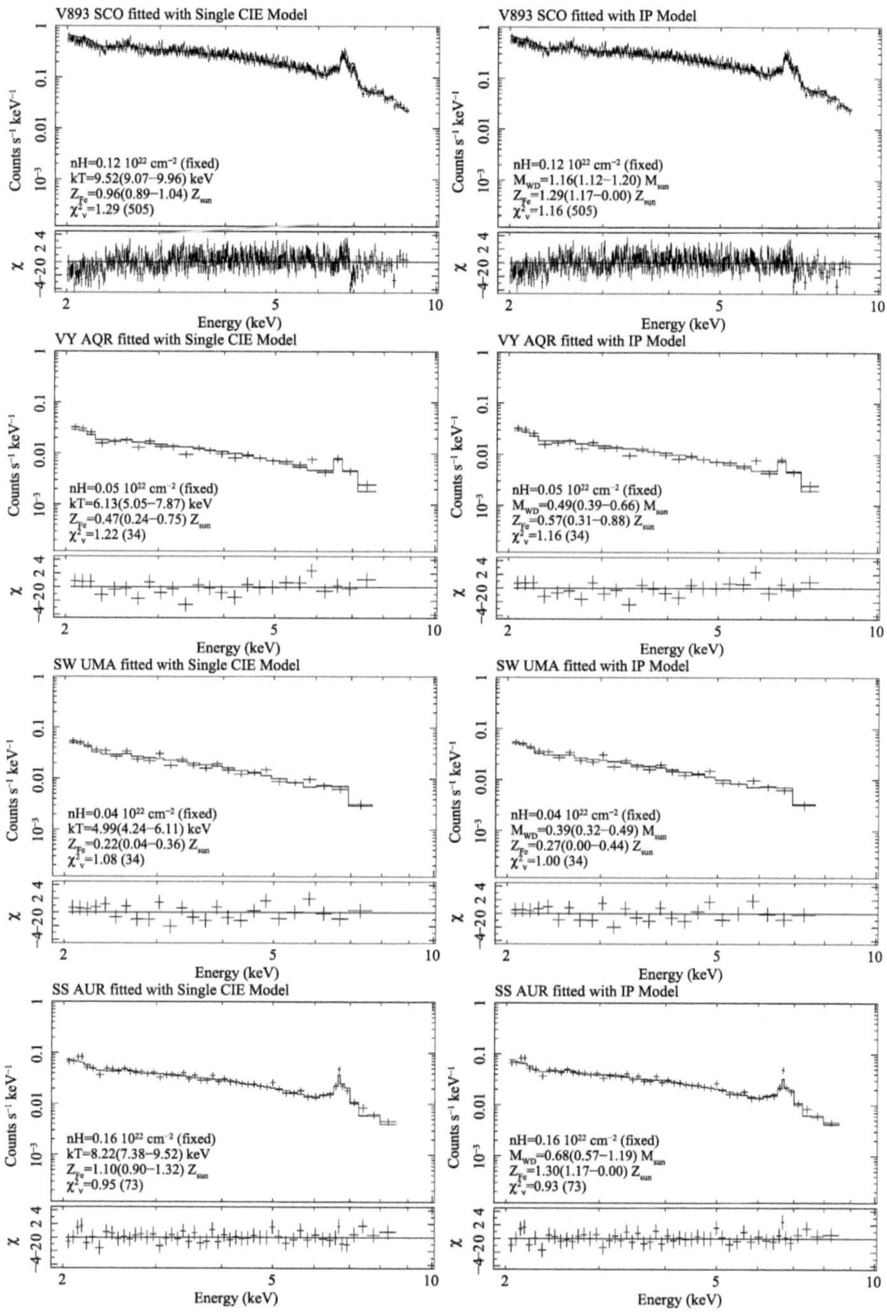

Fig. C.1 X-ray spectra of dwarf novae fitted with a single-temperature CIE plasma model (*left panels*) and the IP model (*right panels*)

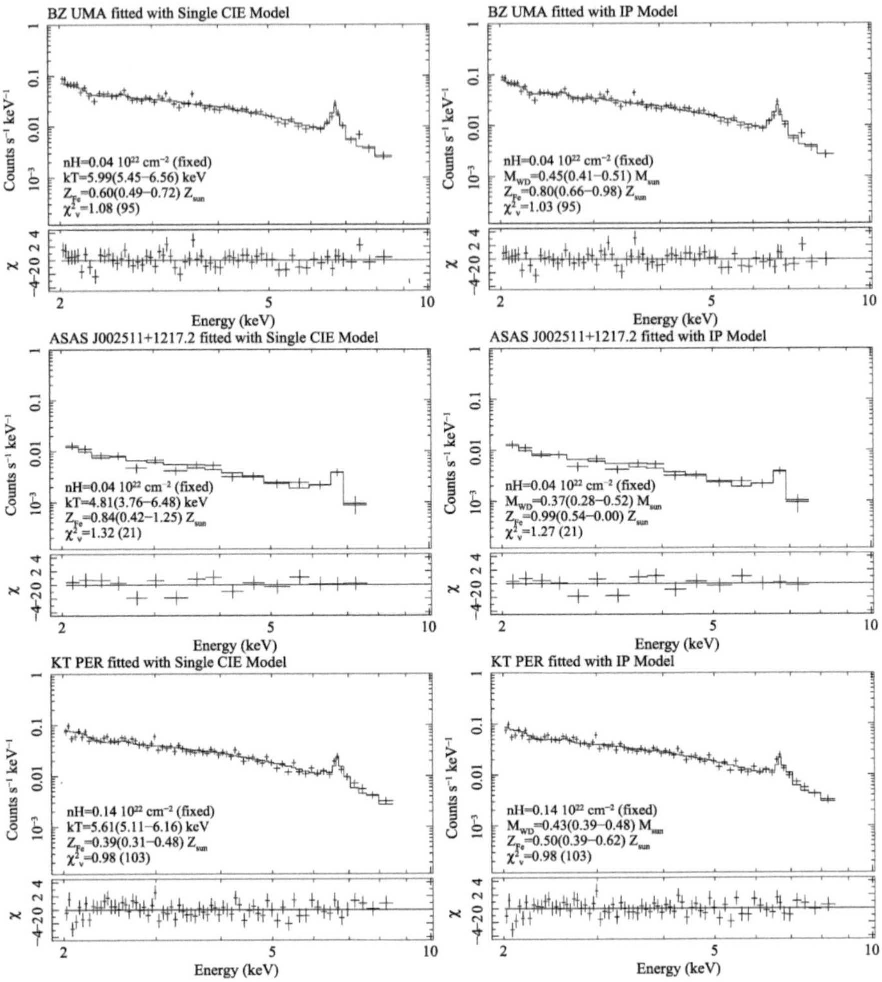

Fig. C.2 Continued from Fig. C.1

References

Kalberla, P. M. W., Burton, W. B., Hartmann, D., et al. 2005, A&A, 440, 775

Pandel, D., Córdova, F. A., Mason, K. O., & Priedhorsky, W. C. 2005, ApJ, 626, 396

Revnivtsev, M., Sazonov, S., Gilfanov, M., Churazov, E., & Sunyaev, R. 2006, A&A, 452, 169

Sazonov, S., Revnivtsev, M., Gilfanov, M., Churazov, E., & Sunyaev, R. 2006, A&A, 450, 117

MIX
Papier aus verantwortungsvollen Quellen
Paper from responsible sources
FSC® C105338

If you have any concerns about our products,
you can contact us on
ProductSafety@springernature.com

In case Publisher is established outside the EU,
the EU authorized representative is:
Springer Nature Customer Service Center GmbH
Europaplatz 3, 69115 Heidelberg, Germany

Printed by Libri Plureos GmbH
in Hamburg, Germany